Scientists and the Regulation of Risk

For Gaspard

Scientists and the Regulation of Risk

Standardising Control

David Demortain

IFRIS, Université Paris-Est Marne-la-Vallée, France and CARR, London School of Economics, UK

Edward Elgar
Cheltenham, UK • Northampton, MA, USA

Published by
Edward Elgar Publishing Limited
The Lypiatts
15 Lansdown Road
Cheltenham
Glos GL50 2JA
UK

Edward Elgar Publishing, Inc.
William Pratt House
9 Dewey Court
Northampton
Massachusetts 01060
USA

A catalogue record for this book
is available from the British Library

Library of Congress Control Number: 2011925756

MIX
Paper from
responsible sources
FSC® C018575

ISBN 978 1 84980 943 6

Typeset by Servis Filmsetting Ltd, Stockport, Cheshire
Printed and bound by MPG Books Group, UK

Contents

Figures

Abbreviations

ADI	Allowable Daily Intake
ADR	Adverse Drug Reaction
AFSSAPS	Agence Française de Sécurité Sanitaire des Produits de Santé
BSE	Bovine Spongiform Encephalopathy
CCP	Critical Control Points
CHMP	Committee for Human Medicinal Products (formerly CPMP)
CPMP	Committee for Proprietary Medicinal Products (now CHMP)
CIOMS	Conference for the International Organization of Medical Sciences
DEFRA	Department for Environment, Food and Rural Affairs
DSUR	Development Safety Update Report
EFPIA	European Federation of Pharmaceutical Industries and Associations
EFSA	European Food Safety Authority
EMA	European Medicines Agency
EMEA	European Medicines Evaluation Agency
EPA	Environmental Protection Agency
EU	European Union
FAO	Food and Agriculture Organization of the United Nations
FCT	Food and Chemical Toxicology
FDA	Food and Drug Administration
HACCP	Hazard Analysis Critical Control Point
HoA	Heads of Agencies
HMA	Heads of Medicines Agencies
ICH	International Conference on Harmonisation
ICMSF	International Commission for the Microbiological Safety of Food
ILSI	International Life Science Institute
INRA	Institut National de la Recherche Agronomique
ISPE	International Society for Pharmacoepidemiology
JECFA	Joint Expert Committee for Food Additives

Scientists and the regulation of risk

MAFF	Ministry of Agriculture, Fisheries and Food
MCA	Medicines Control Agency
MHRA	Medicines and Healthcare products Regulatory Agency
NACMCF	National Advisory Committee on Microbiological Criteria for Foods
NASA	National Aeronautics and Space Administration
NGO	Non-governmental Organization
NHS	National Health Service
NOAEL	No Observed Adverse Effect Level
NRC	National Research Council
OECD	Organisation for Economic Cooperation and Development
PLM	Post-Launch Monitoring
PMM	Post-Market Monitoring
PMS	Post-Marketing Surveillance
PSUR	Periodic Safety Update Report
PVP	PharmacoVigilance Planning
SCF	Scientific Committee for Food
SPC	Summary of Product Characteristics
SPS	Sanitary and Phytosanitary Measures
WHO	World Health Organization
WTO	World Trade Organization

Acknowledgements

Social science can be practised in more or less risky ways. Pursuing one unchanging line of research, whatever the novely of the findings it yields, is one option. It is safer than others, but it is often associated with decreasing returns. Investigating the surprises emerging from the field and using them as a hunch is another option, and often a fruitful one. But conversely there are more risks involved. This book is the result of the second option. All of the findings exposed were unexpected with regard to my initial research project. I intended to investigate the power of independent regulatory agencies, by comparing European food and pharmaceutical agencies. My strategy was to circumvent the usual controversies that seemed to justify the creation of these agencies and determine their success or failure, to study the admittedly more mundane and nitty-gritty aspects: what people in these agencies do, their day-to-day acts and practices, their expertise and ways of thinking, the way in which the objects of their action are constituted, categorised and classified, and so on. One thing made a strong impression on me (and despaired me a little too) as I started to interview and observe people. All used a highly coded language in the most natural way to describe the regulatory activities in the area. They had a language of their own. They employed a number of very neutral and procedural terms, often neologisms, without apparent need to define them even though there was nothing natural in regulation being constructed by such terms. 'Periodic safety update report', 'post-authorisation safety study', 'marketing authorisation' and 'pharmacovigilance', 'risk assessment', 'risk management', 'traceability', 'rapid alert system', 'identification of emerging risks' or 'epidemio-surveillance' sounded to me like a very odd glossary. They indicated that the daily practice of regulation seemed coded and to some extent also codified. There was enough there I thought to depart from my original plans and I set out to understand how this knowledge to conduct regulatory activities comes about.

The result of this intellectual journey is exposed in the book, so I need not go into more details here. But I want to thank the people who took the risk of accompanying me down this road. Jean-Claude Thoenig could have discouraged me from taking such an unprepared change of direction and pushed me on safer paths. Instead, his interest grew as I progressed and he supported this research with his infinite skill, lucidity and motivation for creative research. I am grateful for the chance I have had to work with him. I want to thank Pierre-Benoit Joly as another person who followed my research as it developed and helped it. Numerous exchanges with him contributed to shaping my analysis and this book.

I have made other very inspirational encounters that I would not have had I continued on the original itinerary. I have had the pleasure to work with Mike Power at the LSE Centre for Analysis of Risk and Regulation (CARR) for a few years now. Key ideas in this book have sprung from discussions with him. I can only hope he will appreciate and possibly even profit from what I have done with these conversations. Bridget Hutter was there at important times when further motivation and organisation was needed to continue with this project, distilling important pieces of advice along the way. In this sense she is fully associated with the outcome and I hope this will not displease her.

Other people were essential for writing this book, notably the many scientists who I interviewed and who were extremely generous with their time, information and insights. It often was a pleasure and enriching to meet them, and some indeed became friends on the way. In this context I want to express my gratitude to Peter Arlett, Gaby Danan, Jean-Louis Jouve, Bevan Moseley, René Royer, Michel Van Schothorst, Philippe Verger and Patrick Waller. There are many others who agreed to be interviewed, in too large numbers to be thanked here personally, but I want to express my deep appreciation.

This book would have been totally different if it had not been written in CARR, with its unique blend of people and perspectives. My thanks extend to past and present members of CARR, with whom this work was shared and tested in more or less formal ways: Michael Barzelay, Yasmine Chahed, John Downer, Sarah Dry, Martin Giraudeau, Jeanette Hofmann, Lisa Kurunmaaki, Javier Lezaun, Martin Lodge, Andrea Mennicken, Peter Miller, Yuval Millo, Carl Macrae, Henry Rothstein, Rita Samiolo and Jakob Vestergaard. Outside CARR, I have greatly benefited from the comments of three people in particular: Frank Cochoy, Didier Tabuteau and Peter Weingart. I thank the following for very fruitful discussions on various occasions, including Jean-Claude André (who got me interested in the issue of risk), Daniel Benamouzig, Laure Bonnaud, Olivier Borraz, Soraya Boudia, Steve Collier, Antoine Debure, Jim Dratwa, Matthew Eagleton-Pierce, Jean-Christophe Graz, Boris Hauray, Frederic Keck, Justus Lentsch, Claire Marris, Erik Milstone, Andrei Mogoutov (the man behind the scientometrics used in this book), Mick Moran, Sigrid Quack, Akos Rona-Tas, Stefan Schepers (who co-funded this project), Ebba Sjogren and Didier Torny.

There are a few people who contributed to shape the book as it finally is and I owe them a lot. It has been a pleasure to work with Catherine Elgar and Tim Williams, who showed great enthusiasm for my project and made the whole publishing process as pleasant as it can be. Pauline Khng did a great job copy-editing and improving the manuscript, with great reliability, swiftness and kindness. I had a lot of fun and also learned quite a few things sharing an office and the challenge of getting a book done with Carl and John. I am heavily indebted to the following colleagues who looked at draft versions of all or part of this book: Andrea, Carl, Javier, Jeanette, Mike and Rita. They deserve special thanks here for giving me the opportunity to improve it so much – and to have a glimpse of what academic solidarity and friendship is. I hope they will enjoy rediscovering the book in its final print version and possibly learn something from it for them too. Finally, Charline deserves special thanks for weathering the many ups and downs that inevitably come with this enterprise of writing a book, and specifically for unfailing support during the arduous final weeks of work.

1. Risk regulation – from controversies to common concepts

Scientists shape risk regulation more than they would care to admit. Beyond the assessment of risks, a commonly accepted role of scientists in regulation, they collectively shape standards for managing risks. They might be tempted to deny they have this influence, and in a sense they are right to do so. Only a very specific kind of scientist seems qualified to standardise control: a transnational and multi-professional elite who circulate almost invisibly among the different actors of the regulation of risks. These are people that advise regulators and policy-makers, carry out experiments for and with businesses, and sit in international standard-setting committees alongside their professional practice as researchers, risk assessors or even physicians. Circulation in the different spheres of regulation explains their ability to articulate acceptable regulatory concepts, that is, ideas about the benefits of extending a practice to address a type of risk. This book investigates how invisible colleges of scientists produce such concepts in the domains of medicines safety, food hygiene and novel foods (Chapters 4 to 6). Before this, the relation between science and regulation is explored theoretically and historically (Chapters 2 and 3). The sociological and political implications of scientists' action on regulatory concepts are investigated thereafter (Chapters 7 and 8). The present introduction explains why what is dealt with here is of interest at all and charts the structure of the book.

DISPUTES AND AGREEMENT IN RISK REGULATION

Nowadays, science is seen to pose as many problems as it solves. This perception is strong when it comes to health risks. The era of risk began when we stopped assuming that we could reach a state of absolute safety and reduce risks to zero (Goldstein, 1990; Short, 1992). Since then, we have also stopped trusting scientists completely. What scientists say about the cause and extent of these risks,[1] and the measures to solve them, is almost invariably a source of dispute. This is in spite of the propositions that science has made to solve uncertainty. Science has responded to the lack of knowledge and information about the exact nature, severity and time of occurrence of hazards, with the use of probabilistic calculation. Probabilities have changed the way in which the responsibilities for accidents and contaminations are approached (Ewald, 1986). They are the basis of a new scientific discipline of risk analysis (Starr and Whipple, 1980),[2] which has spread widely to apply to consumer products from medicines, vaccines, medical devices, food supplements, food additives, genetically modified foods to pesticides and veterinary medicines (for the residues they can leave in food). All now undergo a risk evaluation to estimate the severity of a contamination and its causes, limit and doses at which consumers can be safely exposed to a substance or consume a particular ingredient, and so on. These calculations are informed by the idea that uncertainty is best tackled by employing quantitative methods of prediction and probabilistic calculation, based on estimates by experts where data are missing.

Unfortunately, probabilistic thinking is synonymous with politicisation of science – and with disputes. In probabilistic calculation, data gaps and lack of knowledge are compensated by expert assumptions, such as extrapolating from the reactions induced by exposure of a rat to a certain dose of a substance to the reactions caused in humans by the exposure to another dose. These assumptions are almost always influenced by particular epistemic cultures and normative values. They are therefore disputable, and in fact often disputed. Risk analysts recognised early on that values interfere with the choice of assumptions and risk criteria (Raiffa, 1982) and that risk assessment may be partly politicised and conflict-ridden as a result. More than a science, risk assessment is what Alvin Weinberg, a nuclear physicist involved in the Manhattan project and influential commentator on science policy matters, called a 'trans-science': a science that looks at questions that are laid out in scientific language but are 'unanswerable' by it (Weinberg, 1972: 209). He pleaded later for regulators to simply stop asking such questions (Weinberg, 1985).[3] Funtowicz and Ravetz (1992) similarly speak of 'post-normal science' as a science in which the usual heuristics of discovery cannot be applied, given that 'facts are uncertain, values in dispute, stakes high, and decisions urgent' (Funtowicz and Ravetz, 1992: 252). Following Weinberg (1985), Jasanoff (1990a) has defined 'regulatory science' to be a science that is aimed at filling knowledge gaps, producing integrated assessments and predicting the occurrence of adverse health situations. This science is carried out under tight deadlines. It has to face strong commercial, institutional and sometimes public pressures, which open complicated negotiations around the acceptability and credibility of its evidence and claims.

Risk analysts have attempted but failed to maintain the status of science as provider of objective and authoritative advice. Corporatively, they indeed remain realists when it comes to risk (Rushefsky, 1982; Gabe, 1995). They believe in the fact that uncertainty is merely technical: it resides in knowledge gaps that can be filled. Risks are objective facts or calculations that positively capture situations occurring out there, independent from our perception and action on them (Otway and Thomas, 1982; Bradbury, 1989; Renn, 1992, 2008; Rosa, 1998; Hansson, 2010). The solution for avoiding politicisation of science consists in proceduralising decision-making, requiring that all assumptions on which decisions are based must be explicated. Making assumptions and values explicit will help progress towards objective assessments. In this perspective, psychometric analysis serves in reconstructing the assumptions of the public as individual 'risk perceptions'[4] (Fischoff et al., 1978) and objectifying them as parameters of the decision.[5] The main problem in risk regulation supposedly is prioritising the 'right' risks and consistently addressing them (Breyer, 1993; Noll, 1996). Various frameworks have been designed to serve this purpose of proceduralising decision-making, from Lowrance's two-stage process of measurement of risk and judgement of acceptability (Lowrance, 1976) to Rowe's distinction between technical risk estimation and normative risk evaluation (Rowe, 1977). The framework of 'risk analysis', understood to comprise two main sequences of risk assessment and risk management (National Research Council, NRC, 1983; Short, 1984; Pollak, 1995) finally established itself as the standard.[6]

Sociologists and anthropologists of risk however have shown that this ideal of objectivity is unattainable.[7] In matters of risk, dissensus is the rule. There is no reality of risk outside the different definitions, social constructions or selections of these kinds of uncertain adverse events (Douglas and Wildavsky, 1983). Risks are in themselves evaluations and selections of dangers, cognitive operations that reflect particular frames through which social actors interpret the social and physical world (Wynne, 1980; Johnson and Covello, 1987). In this perspective, uncertainty is not just technical but structural or political (Schwarz and Thompson, 1990): it is plagued by the opposition and negotiation between competing claims about what regulatory intervention should be aiming at. In these political contexts, scientists are just one voice, and not a united one. In the words of Beck:

> In definitions of risks *the science's monopoly on rationality is broken*. There are always competing and conflicting claims, interests and viewpoints of the various agents of modernity and affected groups, which are forced together in defining risks in the sense of cause-and-effect, instigator and injured party. There is no expert on risk. (Beck, 1992: 29; emphasis in original)

Science not only fails to avert controversies, but is their very cause. Risks are, following Douglas (1990), like taboos of ancient societies: aspects on which the integrity of a social collective is put at stake and which therefore spark social mobilisation to recall and apply common beliefs and norms. When they touch on these matters of social integrity, they inevitably spark public controversies and disputes (Nelkin, 1984). They come to be feared as much as trusted (Nelkin, 1975). This is especially the case where risk assessment processes are more adversarial, as they are in the USA (Brickman et al., 1985; Jasanoff, 1990b) or where closure around the assumptions and practices of risk assessment experts fails to be achieved (Wynne, 1992; Irwin et al., 1997). In such contexts, science does not speak with one voice, the illusion of apolitical expertise collapses (Benveniste, 1972), experts disagree and conflict in such a way that their assumptions and frames of understanding are revealed, leading to an overall loss of credibility (Weingart, 1999). Science is fully embedded in a form of definitional politics around the problems of the public. This has been extensively illustrated by the study of issues such as nuclear radiation and siting of nuclear plants (Wynne, 1982), the mad cow crisis (Miller, 1999), biotechnological foods (Levidow et al., 2007a) or the location of cell phone masts (Borraz, 2007a), cases in which an often narrow and technical framing of problems as 'risks' in terms of the dominant scientific expertise is slowly displaced by other frames where 'lay people', patients, farmers or social movements in general are aligned.

To sociologists of science, this degree of contestation is an expression of the fundamentally political nature of knowledge production, rather than a new condition of modernity (Beck et al., 1994). Science is shaped by politics in its very act of producing knowledge, as opposed to its being politicised when it gets mobilised in policy-making by non-scientific actors. More than a politicisation of an otherwise pure science, risk assessment is a situation in which the politically and socially constructed nature of scientific assessments is revealed. Most often than not, risks generate radically different interpretations of the same fact, without apparent possibility to agree on any one particular claim (Collins, 1981a). Facts and methods are rigorously deconstructed, and the acceptability and credibility of scientific information negotiated (Jasanoff, 1990a; Wynne, 1992). In many ways, scientific advice and risk assessment form the continuation of politics by other means, something that has been widely confirmed (Wynne, 1992; Cozzens and Woodhouse, 1995) whether in the context of public controversies (Wynne, 1980; Rip, 1986a), judiciary disputes (Wynne, 1989; Jasanoff, 1995a) or scientific advice to government (Salter, 1985, 1988; Jasanoff, 1987, 1990a, 1995b; van Eijndhoven and Groenewegen, 1991).

In sum, the risk society is also a 'controversy society', and science has a lot to do with this. The popularity of this theme however hides another facet of the relation between science and regulation. The fact is, there are many common assumptions about risks in our societies. The long litanies of catastrophes and looming disasters that all too often serve to anchor the risk society argument[8] should not overshadow the fact that risks may also be captured and objectified through 'typifications' (Scott, 1998: 76), which are not much disputed. By this term, Scott refers to the aggregated and stylised facts, expressed in standard language, by which authorities and officials simplify a reality to act more efficiently and directly on it. Scott uses the following example (among many others) that is not unrelated to food and medicine risks (1998: 77):

> The Center for Disease Control in Atlanta is a striking case in point [of typification]. Its network of sample hospitals allowed it to first 'discover' – in the epidemiological sense – such hitherto unknown diseases as toxic shock syndrome, Legionnaire's disease, and AIDS. Stylized facts of this kind are a powerful form of state knowledge, making it possible for officials to intervene early in epidemics, to understand economic trends that greatly affect public welfare, to gauge whether their policies are having the desired effect, and to make policy with many of the crucial facts at hand. These facts permit discriminating interventions, some of which are literally lifesaving.

There are many such typifications concerning food and pharmaceuticals. These include the notion that all medicines are potentially dangerous depending on the dose at which they are taken; that modern medicines are more likely to have side-effects because they are designed to have precise potent therapeutic benefits; that, in general, me-too medicines[9] pose more risks for a negligible added benefit; that the foods which have been consumed as part of a traditional diet can be presumed to be safe; that the side-effects of food additives and other substances in food can be avoided if they are consumed at a small enough dose; that the same specifications can be applied to chemical substances in food that have the same molecular structure because their effects are by definition similar.

All of the above are examples of macro-level evaluations of what is considered 'adverse to health'. They are formulated in broad terms, and focused on classes of events and technologies rather than on individual issues or salient disasters. They are not necessarily quantitative as would be expected of a typical risk assessment. They constitute a sort of pre-regulatory knowledge. They circulate in professional and scientific communities, inform the discourse of regulators and politicians, and may indeed become the foundation for popular conceptions of risk. These assumptions are rarely disputed. At the very least, they often remain unquestioned for long periods of time during which they justify deploying these activities of control that constitute risk regulation.

Risk regulation is not a domain as such (Hood et al., 2001),[10] but a set of activities for attempting to control uncertain adverse health events. To establish an official product specification (e.g. define the dose at which a particular substance present in food can be consumed), a whole set of standardised scientific activities need to be conducted. These include rearing of laboratory animals, development of materials and devices for chemical analysis, development, validation and standardisation of test protocols, conducting of tests, experimental research, programming of research to fill knowledge gaps, launch of risk surveys to gather more data and inform further risk assessments, sampling and analysis to know at what dose this substance may ultimately be found in food, and so on. Products are tested and labelled, standards are set and enforced, masses of data are collected and analysed, inspections are carried out daily to sample and check compliance with specifications. Risk assessments are produced every day by regulatory agencies and experts, with all of these activities occupying a huge number of personnel in government at all levels, as well as in industry. These daily routine activities can only continue because they are founded on a minimal agreement on what to target in regulatory intervention and how to go about it. There are accepted criteria to define which adverse events are worth regulating. This is the role of the typifications that Scott speaks about. It has three other characteristics that deserve emphasis.

First, these standard evaluations of risk emerge thanks to a material infrastructure of tools and operations by which events and products are traced and observed, and information is coded and reported as well as analysed. This infrastructure characterises the problems that should be corrected and form the purpose of regulation as a whole. It makes 'information gathering' possibly the most important tool of the government of risk (Hood, 1983; Hood et al., 2001) by anchoring governance in networks of observation and experimentation (Sabel and Zeitlin, 2010). These standard evaluations emerge along with epidemiological thinking and practice, with the ambition to construct new realities out of the statistical observation of regularities (Desrosières, 2002).[11] Systems of surveillance, information-gathering and interpretation that allow the production of these standardised evaluations have grown in recent decades. There is hardly a product or industrial activity that is not included in an infrastructure of observation, sampling and reporting of information. As far as food and pharmaceuticals are concerned, all adverse drug reactions (ADRs) by law should be signalled, in particular serious and rare ones. Dedicated formats, reporting lines and central databases have been set up to collect and analyse them nationally and internationally. This means that between 2001 and 2004, more than one million reports on adverse drug reactions were collected by national, European and worldwide centres of pharmacovigilance such as the World Health Organization (WHO) centre for the monitoring of medicines.[12] In food, a company is required to record and signal contaminations for inspectors to track. Veterinarians and physicians are also required to notify certain diseases that are carried by food, and networks of epidemiological surveillance, in which these professionals enter common observed cases of disease to assess their prevalence, have grown in numbers in the last three decades.

Second, risk typifications result from a search for responsibility and aim to reform intervention. Risks are evaluated for the probability that they will occur and to define the ways in which they may be averted. We do so by the analysis of human decisions that could cause them (Luhmann, 1993) with the ambition of proposing actions to avert them. As social constructivists say, risks 'do not exist out there' but are always the product of a particular way of approaching the event, of a mode of recognition by which potential events are brought into being in the present (Ewald, 1986; Hilgartner, 1992; Garland, 2003). Evaluations are also linked to intervention since by necessity they involve an evaluation of these assumptions and criteria as much as of the event itself, in particular when negative incidents – or outright disasters – materialise that prove these assumptions wrong. A risk evaluation is thus always a regulatory evaluation.[13] For instance, risk assessors when considering the data about the number of accidents linked to the exposure to a substance are always simultaneously in a position to state whether pre-existing measures to reduce this exposure were effective or not.

Third, a characteristic of these typifications is that they are the translation of a form of intervention: the application of certain assumptions and criteria – some may say moral views – that define specifically what to look for and what information to collect. Lezaun (2006) thus shows the work involved in establishing an agreed definition of what a genetically modified food is – more specifically the 'transformation event' that qualifies it as modified, and the ensuing installation of devices that specifically track this transformation and the organism as a whole. Typifications thus emerge because there is coverage of a wide territory of similar events and because of the intention to spread and standardise intervention over it. As hinted by Scott (1998), using Porter (1995) to entrench this idea, typifications do not emerge from the willingness to construct an objective and precise image of what is happening out there, but as an attempt to defend this intervention and standardise it. Adverse events are characterised and anticipated in such a way that they ultimately justify extending an intervention to these phenomena.[14]

Arguably then, and even in disputed matters of risk, knowledge is shared that defines the generic issues to be addressed and indicates which potential effective activity is best to counter the problems. This shared knowledge is embodied in what we may call regulatory concepts or abstract ideas about the benefits of standardising a practice to address a type of risk.[15] Several are studied here: the proposition that we would be better prepared to act when unexpected adverse drug reactions occur, and to mitigate them if all that is known about a drug is recapitulated and centralised in one place before it enters the market; the notion that the prevalence of food contaminations could greatly be reduced if all food businesses defined in advance what are the most likely contaminations for their products and monitored them as foods are produced, instead of checking if they have occurred after production; or again that methodologies to survey consumptions and side-effects of novel foods may help to bring reassurance that anticipated risks will not materialise. These concepts contain more or less implicit theories for regulatory action (Sabatier and Hunter, 1989; Hofmann, 1995) in the form of causal narratives – 'generalising practice A will impact and reduce effectively problem B, provided conditions X and Y are fulfilled'. They are generally accepted as valid and beneficial, which is why they become authoritative and regular (Drori et al., 2003) – in effect they are standards.

SCIENTISTS AND THE STANDARDISATION OF CONTROL

What role do scientists play in the emergence of these concepts and why should they be singled out as standardising actors? From state intervention through command-and-control, risk regulation has become much more of a transnational system involving all sorts of decentred mechanisms and actors that contribute to exert control over private activities (Baldwin et al., 1998; Baldwin and Cave, 1999; Hutter, 2001; Black, 2002). In contrast with Selznick's oft-quoted definition of regulation, focusing on public authorities,[16] regulation involves multiple private and non-state actors that constitute, along with governmental institutions, the 'regulatory space' (Hancher and Moran, 1989). Regulation is not any more just an action of courts. There are many more regulators. In food and pharmaceutical regulation, for instance, companies are turned into a kind of public actor, being embedded in a 'web of controls' (Braithwaite, 1993; Braithwaite and Drahos, 2000) by which they are led, for instance, to signal the problems arising with their own products or activity and share them with inspectors or regulatory agencies. They mobilise their capacities and expertise to internalise and respond to regulatory demands and preoccupations (Coglianese and Lazer, 2003; Power, 2007). They contribute to set, implement and monitor standards concerning the risks themselves or, more and more in fact, the ways to define and address them.

Risk regulation also increasingly takes place transnationally. The category of the transnational serves first and foremost to avoid opposing the local and the global, or the national and the international, as being two different 'scales' (and by the same token to avoid searching for which scale matters more than the other, or whether the state does or does not decline in a context of globalisation). It simply recalls that whatever is happening in one national context is dependent on or linked to what is happening in another and that some actors and institutions tend to embody and act on this interdependence. Djelic and Salhin-Andersson are right to observe that:

> The label 'transnational' suggests entanglement and blurred boundaries to a degree that the term 'global' could not. In our contemporary world, it becomes increasingly difficult to separate what takes place within national boundaries and what takes place across and beyond nations. The neat opposition between 'globalization' and 'nations', often just beneath the surface in a number of debates, does not really make sense whether empirically or analytically. Organizations, activities and individuals constantly span multiple levels, rendering obsolete older lines of demarcation. (Djelic and Salhin-Andersson, 2006: 4)

A focus on the transnational also contrasts with an approach in terms of 'transboundary risk management' (Linnerooth-Bayer et al., 2001) which reduces the transnational to the interactions between state or governmental authorities to improve the management of risks that tangibly cross borders. Of course, regulation continues to be influenced by national regulatory styles or political cultures in important ways (Vogel, 1986; Rothstein, et al., 1999; Jasanoff, 2005). But the notion of the transnational is here to recall that actors and rules, even if placed in a local or national context, are always interdependent with those emerging in other geographical and political locales. These chains of interdependence do not necessarily extend over the globe: the transnational is not the global. But in turn they accentuate the interplay between intervening actors, private and public, state and non-state (Djelic and Sahlin-Andersson, 2006; Black, 2008; Botzem and Hofmann, 2010), notably as international standard-setting bodies and various clubs, networks or councils of various sorts are established that help professional and industrial actors to turn themselves into producers of standards in their own right (Prakash and Potoski, 2006).

While this book is partly about the setting of standards, the focus is on the processes that precede these formal acts. I do not exclusively concentrate on the production of formal rules, as initially studied by political science (Mattli, 2001; Mattli and Büthe, 2003) and by the SCORE school (Brunsson and Jacobsson, 2000), but on standardisation as the alignment of actors of a regulatory domain on similar ideas about how to control uncertain adverse events and with what effect.[17]

Stating that scientists standardise control of products means that, beyond their professional role of testing products and substances, assessing risks or monitoring adverse drug reactions or food illnesses, they act among and between these various actors to articulate common concepts. This comes in support for the argument of neo-institutionalist scholars that scientists are among the major standardisers of this world (Brunsson and Jacobsson, 2000; Jacobsson, 2000). As holders of an expertise, they provide international standard-setting bodies with the legitimacy to set rules even in the absence of power of enforcement or sanction (Hülsse and Kerwer, 2007). They benefit from the cultural legitimacy accruing to science, to develop and carry universal and rational scripts for organisation and action (Drori et al., 2003; Drori and Meyer, 2006). This is happening globally, and in more and more organisational domains. It affects administrative processes and activities of governance (Drori et al., 2006). Regulatory concepts as world models replace states as vectors of rationalisation, in a decentralised and fragmented world polity (Meyer et al., 1997; Boli and Thomas, 1997). Scientists and professionals are among the main carriers of world models (Meyer, 1994), the vectors by means of which these ideas are conveyed and translated (Czarniawska and Joerges, 1996; Engwall and Salhin-Andersson, 2002).

But how does such science-driven standardisation come about? What explains the capacity of scientists to produce and standardise concepts for the control of health risks? How is the setting of international standards connected to scientists' basic professional activities of research and risk assessment? Where do regulatory concepts come from and how do they gain relevance and legitimacy as modes of organising activities of control? Scientists are just one actor of transnational regulation, but they have a distinct capacity to foster agreement around concepts of control. This is because they work, individually and collectively, across regulatory domains, developing an ability to integrate them and formulate, as well as take responsibility for, their outcomes. Their concepts are recognised by the different types of actors of a regulatory domain – industry, professionals and public authorities. In fact, one could say that they break the boundaries separating these actors and show the irrelevance of these categories (Ong and Collier, 2005). Again, not all scientists do this. Those who do are polyvalent and influential scientists who circulate among the different spheres and actors of control, and do so transnationally – invisible colleges. There are often several of these scientists in the same area. They may or may not know each other, but they exchange elements of practical experience of risk and control as well as ideas, and at times get together to agree on common concepts. This invisible circulation explains the articulation of regulatory concepts and their editing as formal, harmonised rules. Concepts create the common ground for the adoption of formal standards, the cognitions that explain that 'consensus' can be produced within an international standard-setting body. These particular moments when loosely tied colleges of expert scientists make the junction between evaluations of risk and control on the one hand and international setting of standards on the other are what this book sets out to explore and analyse to fully appreciate the contribution of scientists to the transnational regulation of food and medicines risks.

ORGANISATION OF THE BOOK

The book is structured around the study of three regulatory concepts that concern the safety of foodstuffs and pharmaceuticals. Before getting to those case-studies, Chapter 2 presents the benefits of approaching scientists and their evaluative activity through the concept of invisible colleges, as opposed to other available notions such as that of epistemic community or knowledge networks. There are three main benefits: the emphasis on this characteristic of certain scientists not just to possess social authority and credibility, but to circulate and act transversally thanks to a variety of identities that they can use – researchers, scientific advisers, industry consultants, standard-setters and so on. This multi-professionalism is not simply a smart way of swapping hats to discreetly accumulate more influence. It is inseparable from a capacity to move across the contexts in which adverse events and regulatory activities can be experienced and evaluated. Scientists circulate across a variety of sites – public and industry laboratories, food or pharmaceutical manufacturers, scientific advisory committees, regulatory agencies, national, European and global standard-setting and policy-making bodies–and this is the basis of their ability to standardise. The second benefit of this notion is to emphasise the fact that such scientists are special ones, an elite really. Finally, the notion of invisible college emphasises confidentiality, lack of publicity or in perhaps less normative terms, interstitiality. Scientists of this sort meet and work together in organisations that stand in between rather than confront vested interests and established institutions.

Chapter 3 provides more empirical background for the study of the three concepts. It draws the landscape of food and pharmaceutical regulation in the European Union (EU) and starts documenting the close affinities that exist between these regimes and the research programmes of evaluative sciences such as clinical pharmacology, toxicology or microbiology. These sciences are founded on a historically growing interest in the evaluation of risks, and the belief that uncertainty is best averted if products, substances and ingredients are ranked or classified according to the probability that they cause adverse side-effects. They have given rise, alone or in relation to other disciplines, to the practice of drug safety evaluation, of food control and food risk assessment. These sciences are dynamic: from an initial programme of qualification – the exhausting of the properties of a product, including 'side-effects', as tested and known at a given moment in time – some scientists have evolved towards imputation, which involve ubiquitous attention to emergent signals of adverse events, wherever they occur. Imputation approaches adverse events with full recognition of their complex trajectories, their uncertain and open-ended nature. It searches for the cause(s) – and responsibilities – of such events. Chapter 3 looks at this shift from qualification to imputation through a bibliometric co-word study of these fields. It shows that this shift is reflected in changes to regulation regimes, through the emergence of systems for the monitoring of risks, such as pharmacovigilance, epidemio-surveillance and food surveillance plans. Chapter 3 thus advances the hypothesis that there are sciences that standardise as a result of their interest in the evaluation of product-related health risks. Chapters 4 to 6 test this hypothesis through three cases, each of which covers a specific policy area: the regulation of medicines through the case of pharmacovigilance planning (PVP), the regulation of food hygiene through the case of Hazard Analysis Critical Control Point (HACCP) and finally the domain of novel food regulation approached through the standard of post-market monitoring (PMM).

According to the concept of pharmacovigilance specification and planning or PVP, unexpected adverse drug reactions and patients' exposure to them are better managed if manufacturers and regulatory agencies are prepared for their appearance. Recapitulating what is known but also what is not known about the safety of a drug as it is introduced in the market, and planning the concrete responses to deploy if adverse reactions are discovered, would help make more adapted decisions concerning the presence of the drug on the market and avoid radical measures of withdrawal that affect both the patients that need the drugs as well as the industry. PVP draws the lessons from the lack of communication between manufacturers and regulatory agencies, as well as between the pre-marketing and post-marketing teams in these agencies, and the ensuing inability to tailor measures of risk communication and management. The HACCP concept is a set of abstract principles that defines how food businesses should go about establishing a safety assurance plan that systematically reduces the risks of contamination of their products, during the production or preparation process and beyond. It is grounded in the assumption that risks of contaminations and poisonings will be reduced if individual food businesses adopt a plan whereby the process points at which contamination is most likely to occur are monitored with pre-defined actions to correct the problems should they occur. The HACCP concept builds on the idea that microbiological testing of foods at the end of the production chain is less efficient in mitigating food contamination risks and poisoning because the latter results from lack of caution and competence in the production or preparation of foodstuffs. PMM, finally, is a set of methodologies to collect information about the product's use pattern, consumer profiles and possible association with adverse health events after the launch of a food product. As a concept, it argues that such practices should not be generalised to all novel foods to avoid giving weight to the impression that novel foods, like medicines, systematically carry side-effects that would be hard to detect in pre-marketing tests.

All three concepts advocate the monitoring of risks and contain complex protocols by which this can be carried out. They refer to experimental practices in the pharmaceutical or food industry, deployed and evaluated with the support of academics or researchers who were at the same time or later in their career connected with regulatory agencies. As a consequence of this scientific supervision, these experiments (often as yet unnamed though they would be eventually) are discussed in academic publications, and later come to be discussed in official meetings of international policy-making and standard-setting bodies and integrated in European legislation.[18] All three concepts represent a synthesis of similar evaluations emerging in different places at the same time. It is possible to trace the links between these experiments and discussions, by which a common formula displaces previous ones to create the impression, in hindsight, that the concept has always existed or has diffused from one clear point. The chapters thus describe the colleges by which these concepts crystallised, the mode of theorisation of control – lesson-drawing, abstraction, modelling – that was used and the process by which these concepts were translated into formal international standards.

There are of course important differences between those three cases, notably as concerns the junction between evaluation and standard-setting – its speed, its consequences and the type of invisible college that allowed it. PVP crystallised around the year 2000 to become an official international standard as early as in 2005. This concept of planning applies to a great variety of medicines, and inspires a whole set of internal processes in the industry and regulatory agencies, many of them setting up risk management services and new routines of collection, assessment and reporting of safety data emerging from clinical trials. The case of PVP exemplifies a great power to reform regulatory intervention through concepts that are general in scope as well as highly prescriptive. This crystallisation of such impactful concepts can be traced to the existence of a college of drug safety experts who translated a rich collective experience of post-marketing surveillance into a standard, via their direct and active participation in international consensus conferences and standard-setting bodies.

HACCP has a much longer history, extending between the testing of an in-line safety assurance plan by the American company Pillsbury at the end of the 1960s, to the adoption of an international guideline by the Codex Alimentarius in the mid-1990s and a European Union Regulation mandating its application in 2006. Much like PVP, HACCP has a very wide scope of application. It was immediately promoted by international organisations such as the WHO as applicable in all countries, all segments of the food industry and all types of foods. But HACCP produces its effects via the mediation of more precise and concrete guidelines, designed for various segments of the food industry by national governmental or professional bodies. The concept came to exist and became a standard by the mediation of elite food microbiologists who modelled a relatively limited practical experience – focused on microbiological issues of food safety in large production-line food industries – to respond to high-level and global policy challenges of international bodies like the World Health Organization and overcome the fragmentation of food hygiene practices and norms.

PMM offers a striking contrast with PVP and HACCP. PMM was specifically defined in a paper published by a group of food risk assessors gathered together by an industry-sponsored scientific foundation. This was done on the basis of a restrictive regulatory concept, according to which generalising PVP would accreditate the view that novel foods pose a general safety issue, a standard evaluation that the concerned situations are reluctant to make. The formal criteria for using PMM were therefore deliberately restricted to those situations in which it had already been employed with utility, avoiding its generalisation. The college that articulated the concept and made the junction between evaluation and standard-setting was in this case a small college of risk assessors, and conscious of the limited capacity they had to promote a form of intervention given they had limited experience of novel foods and legitimacy as actors of control and risk management.

These cases pose two further questions that the last two chapters address. Chapter 7 seeks to appreciate in greater detail who these scientists that make the junction between evaluation and standard-setting are, and specifically how they can be differentiated from the broader mass of scientists researching and evaluating risks. It presents data concerning the scientists that were involved in the development of the three concepts, to assert one main thing: they are specialists in evaluative sciences who have moved away from their original activity as researcher or physician and have throughout their careers worked with a variety of regulatory actors, in a way that is self-reinforcing as these careers become more transnational. It shows another thing: engagement in theorisation of regulatory intervention and in the setting of standards is not so much the result of a strategic entrepreneurial attitude towards policy and regulation as it is an opportunistic and short-lived mobilisation through interstitial organisations – consensus conferences, learning societies, foundations – that allow scientists to experience their political capacities without distancing themselves from their professional identity of researcher or risk assessor and the social authority attached to this.

The final chapter sets the focus on the dissimilarities between the three cases, and especially on the differentiated capacity of invisible colleges to act on the whole domain, the functioning of the system of actors and tools that organise the control of products and adverse events. PVP acts on such a system, integrating post-marketing safety and licensing of drugs in a common scheme of risk management, redefining identities and protocols of cooperation in this system. HACCP and PMM – for different reasons – less so. The last chapter correlates this with the capacity and willingness of scientists to take responsibility for the successes and most importantly the failures of a domain. Impact is strong when invisible colleges are fully embedded in regulatory operations and when, by the virtue of a sort of responsiveness to the problems and failures of regulatory intervention as a whole, they legitimately embrace the ambition to reform it. Invisible colleges therefore do not change regulation from the outside. They are emergent from domains, especially from domains that are characterised by high ambitions of integrated control and reliability. It is in such contexts, in which risks of regulatory failure are stronger, that invisible colleges are most likely to emerge and act to settle the responsibilities of everyone, including themselves, in an attempt to satisfy mixed professional and political ambitions to improve the control of risks.

NOTES

1. Throughout this book, I minimise the use of the term 'risk' because of its ambiguity and to avoid confusion. The term refers both to events that have taken place and to future ones. It is sometimes a synonym for the mundane notion of 'potential problem', but also that of 'probabilistic calculation' in other contexts. This shows that when speaking of risk, different modes of perceiving the event may be involved. Speaking only about 'risk' conceals this diversity, and forces one to descend to the level of more neutral categories of 'adverse drug reactions', 'side-effects', 'food contaminations', 'poisonings' or 'food-borne diseases'. For want of a better term, I will also be using the briefer generic term of (uncertain) 'adverse health event' or 'adverse event'.

2. The Society for Risk Analysis was officially formed in 1983 and the first issue of its official journal *Risk Analysis* appeared in March 1981 (Thompson et al., 2005). Health risk analysis emerged in the early 1980s marrying toxicological and ecotoxicological methods with epidemiology and environmental health expertise on the one hand, and probabilistic methods applied in engineering sciences to the control and insurance of physical and chemical hazards on the other. It also takes inspiration from financial risk and cost–benefit analysis in its ambition to establish and apply quantitative criteria for what is too risky or costly at a societal level (Starr and Whipple, 1984; Bernstein, 1995). The medical sciences have a risk tradition of their own (see Hayes, 1992; Skolbekken, 1995; Schlich, 2004). Generically however, the discipline of risk analysis is a strategic science that is concerned with decision-making (Rip, 1986b), which may be observed in the fact that it almost immediately found an application in the processes of regulatory agencies like the Food and Drug Administration (FDA) or the Environmental Protection Agency (EPA) in the USA (Morrison, 1982; Jasanoff, 1992).

3. Quite an early plea, judging from the contemporary attempts in food safety analysis for instance, to ensure that decision-makers take on the task to better define the terms of the question addressed to scientists as well as the rationale for asking it (Millstone, 2007).

4. With a choice of term that marks the asymmetry of power between citizens and experts, with the latter 'assessing' rather than 'perceiving' the risks as ordinary people do (Perrow, 1982).

5. It is therefore an integral part of the discipline of risk analysis and builds on the same premises. Starr's initial insight that risks are over-estimated by individuals (in comparison with statistical frequency) if they feel they do not have control over their exposition to them (Starr, 1969) in fact opened this field.

6. By classifying the operations of decision-making under these two complementary and almost 'indistinct' categories (Short, 1992: 6) this framework brings about the admission that science and politics, or calculations and value pronouncements, are fully entangled though in practice they should be separated (Ashford, 1984). It evolved later on that the public were integrated in the framework through the rubric of risk communication (NRC, 1989) and a recasting of risk characterisation – the intermediary sequence between risk assessment and risk management – as a decision-driven and fully political operation (NRC, 1996).

7. Sociologists approached the matter of risk slightly later than risk analysis practitioners and with a variety of perspectives, sometimes rather close to technical risk analysis (Renn, 1992). In spite of early calls for sociologists to tackle issues of risk analysis (Short, 1984), they have not managed to counter the continued objectification of risk by technical risk analysis (Burgess, 2006). But nonetheless sociologists have proposed elaborating procedures and designing institutions for collective deliberation on technologies and risks (Latour, 1999; Renn, 2004 and 2008; Callon et al., 2009).

8. There are innumerable examples of this. For the sake of illustration, see the introduction to Giddens' 1999 paper: 'What do the following have in common: BSE; the troubles at Lloyds; the Nick Leeson affair; global warming; drinking red wine; declining sperm counts? All reflect a vast swathe of change affecting our lives today' (Giddens, 1999: 1).

9. Me-too medicines are drugs with similar pharmacological and therapeutic properties to already approved drugs, which are developed by the pharmaceutical companies to take their share of the profit generated by a class of product that is known to be profitable. Statins (cholesterol-lowering products) are an example. Products belonging to this class include atorvastatin (Lipitor), cerivastatin (Baycol; see Chapter 4), fluvastatin (Lescol), lovastatin (Mevacor), pravastatin (Pravachol), rosuvastatin (Crestor) and simvastatin (Zocor), developed by different companies and authorised at different times.

10. An initial and somewhat 'unsophisticated' approach to risk regulation (Hutter, 2006: 205) saw it as a form of control of accidents in the name of the safety of workers or con-sumers exposed to health hazards in the course of production or consumption. It was then considered to be a part of social or 'protective' regulation, emerging in the 1960s and 1970s to address imperfections in liability law and other institutions for litigation (Bardach, 1989). The study of the language of risk by Ewald (1986, 1991) showed how this probabilistic and scientifically framed search for causes (instead of moral and crimi-nal responsibilities) was a generic one that affected more and more domains. Later in the 1980s, risk regulation was taken to cover mostly the regulation of health and envi-ronmental adverse events (Hutter, 2006). But just as risk can be defined in multiple ways to form a sort of archipelago of related notions and issues (Hood and Jones, 1996), risk regulation straddles many different domains and captures certain commonalities, such as the increasing importance of frameworks of risk analysis and risk management for organising regulatory intervention.

11. Chapter 3 returns to the importance of statistical and epidemiological thinking in the rise of a research programme of imputation.

12. Pharmacovigilance is now defined as 'the science and activities relating to the detection, assessment, understanding and prevention of drug adverse effects or any other drug-related problem' (WHO, 2002). It is also conventionally based on two different tools: systems of notification of suspected cases by the industry, physicians and other health professionals and regulatory agencies, and pharmacoepidemiological studies looking at the occurrence of side-effects in a large and controlled population of patients using a given drug (or combination of drugs).

13. Although much work remains to be done in this area, it appears that risk analysis and policy analysis are sister disciplines, emerging together and producing similar sorts of tools (e.g. risk assessment and policy evaluation). Both of these disciplines are linked to systems analysis and the ambition to describe the world as systems of coherent ele-ments, from which objective conditions of success or failure of intervention can be derived.

14. Much like the concept of risk itself serves the regulatory state in responding to failures, negotiating accountability and eventually governing expanding areas with minimal costs (Majone, 1994; Clark, 2000; Moran, 2002; Power, 2004 and 2007; Rothstein et al., 2006).

15. Carpenter develops a comparable notion of conceptual power (Carpenter, 2010). My own perspective of regulatory concepts, as a form of shared regulatory knowledge edited under the form of standards, criteria and protocols of control, rejoins his inter-est in the 'formal and informal definition of concepts, vocabularies, measurements, and standards' and attached 'patterns of thought, communication, and learning' (Carpenter, 2010: 64). Carpenter's specific interest is in the fact that conceptual power is located in an organisation, the Food and Drug Administration (FDA), while I see it as distributed in circuits of risk evaluation and colleges of standardisation, most likely because I approach the topic from a transnational perspective and from Europe where regulatory agencies are traditionally weaker than the FDA.

16. The definition is regulation as a 'sustained and focused control exercised by a public agency over activities which are valued by a community' (Selznick, 1985: 363). It should be added that other elements of the definition ('sustained', 'focused', socially valued activities) are as important as the reference to a public agency, even if less often mentioned.

17. This shift in considering standardisation as cognitive alignment (see also Porter, 1992) in a regulatory domain is motivated by the difficulty in defining standards in the way Brunsson and Jacobsson (2000) do, that is as voluntary regulatory rules. In their approach, standards are instruments of control of organisational fields that, by the expertise they incorporate, generate a willingness on behalf of potential users to comply even in the absence of structures of enforcement. Their voluntary nature distinguishes them from 'social norms' regulating social communities rather than organisational fields, and 'legal directives' that are mandatory. This distinction has been criticised (see for instance Dudouet et al., 2006), as having the notions of 'hard' and 'soft law' (Power, 2007). It contradicts the argument of socio-legal studies and sociology of law inherited from Max Weber, that legal rules are only effective in so far as social actors make sense of them in action. It is therefore misplaced to characterise voluntary standards in opposition to imperative laws, because in both cases effectiveness depends on internalisation of the rule. Standards have institutions for enforcement, if decentralised ones (Kerwer, 2005), and thus should be equated with law rather than convention in the Weberian sense (Weber, 1956). What is more, standards are often mentioned in law. However, a defining attribute of formal standards is that they only produce an effect when the user possesses the knowledge and competence to interpret and apply them. Standards create a kind of disarray for users who do not know where the rules come from, what precise needs justified their creation and what situation they are meant to solve in practice. It is possible to argue that formal standards are created after knowledge has been shared and disseminated concerning the content and effects of the practice.
18. While I am conscious of the fact that international standards are applied in different ways in different regions of the world and countries, a focus on one particular region seemed necessary to be able to evaluate the impact of regulatory concepts, which is what European legislative and policy developments are used for here.

2. Communities, networks and colleges: expert collectives in transnational regulation

The relation between science and policy, those who know and those who govern, is an old theme in social science, from politicians' and scientists' distinct ethics and logic of social action (Weber, 1919) to the concentration of scientific resources within overlapping elite political, corporate and military circles (Mills, 1956), passing by the policy-driving role of science-policy institutions and 'scientific estate' in post-World War II USA (Gilpin and Wright, 1964; Price, 1965; Haberer, 1969). More recently, references to various types of transnational, international or even global communities and networks have flourished. Two of these seem to specifically account for the action of scientific professionals in transnational regulation and combine a political and epistemic point of view on these actors, appropriate to understand their standardising action: epistemic communities and global knowledge networks. My intention in this chapter is to present the analytical tools that have been used to account for the hybrid form that transnational expert collectives take, and assess whether and how to use them. Communities and networks are often presented as relational logics that are intermediary between hierarchies and markets, and as such they have been extensively applied to experts to account for the way in which they mediate other actors or organizations (Mayntz, 2010). I will describe what properties of expert collectives they emphasize or on the contrary overlook, to finally argue in favour of a notion of invisible college that is, I think, more appropriate to capture the complex and mixed webs of relationships by which standard knowledge of risks and effective interventions crystallise.

EPISTEMIC COMMUNITIES AND KNOWLEDGE NETWORKS

The concept of epistemic community denotes the way in which a community of people who share knowledge influence the formulation of policy problems and policy design. In his highly cited paper, Peter Haas (1992: 3) remarkably synthesised what an epistemic community is, defining it as 'a network of professionals with recognized expertise and competence in a particular domain and an authoritative claim to policy-relevant knowledge within that domain or issue-area'. Originally, the notion refers to the sharing of knowledge or of an epistemology by a group of people, such as scientists held to be the experts of a domain (Haas, 1989). It has been extended to policy-making situations in the recognition that even in these situations where communities comprise people with different backgrounds, organisational affiliations or even interests, beliefs and knowledge may indeed be shared. There are four characteristics that members of epistemic communities share: normative and principled beliefs, causal beliefs, notions of validity, and a policy enterprise. Experts are influential because they share norms and beliefs, and carry these into policy-making in a coordinated manner as advisers to bureaucracies and politicians. The epistemic community approach locates politics in the decision-making process by which politicians and state administrators make choices concerning policy action, under the cognitive influence of particular ideas and beliefs that affect the preferences of the former.

The concept is used in international relations and in research on national and European policies (on the latter, see for instance Richardson, 1996; Verdun, 1999; Zito, 2001). It is also used by Braithwaite and Drahos (2000: 73–75) to characterise a type of actor of global regulation (alongside states, companies, international business organisations and non-governmental organizations, NGOs). Most papers on epistemic communities cite Haas' concept (1992). His is a compelling one because it has three different layers of explanation. The first is the authority of science. Even though epistemic communities include people other than scientists, the latters' influence is still dependent on them embodying scientific method, prestige and authority (Drake and Nicolaïdis, 1992; Haas, 1992). A related factor is the uncertainty that surrounds certain policy problems, global ones in particular, in terms of the lack of knowledge about their causes as well as possible solutions. Epistemic communities provide the knowledge to compensate this uncertainty. The second dimension of the concept is the community. Policy-makers and regulators from various countries may agree on what policies to implement because they converge around similar expectations about the problems and solutions. They share this knowledge because they are advised by or work with people who belong to the same community of thought. Science, because it is organised in communities or disciplines, increases coordination between policy-makers, even where they have opposing interests. Communities circumvent the politics of interests and power plays, which is a factor for the convergence around common expectations and values (Keohane and Nye, 1977). The third element is the heavy presence and involvement of professionals in politics and public affairs. The concept is used quite broadly to analyse these policy situations in which there is a density of experts advising various parties to the policy-making process. The concept resonates with the literature that shows how the codification and authority of specialised expert knowledge has been a contributory factor in the emergence of public bureaucracies since the end of the nineteenth century (MacLeod, 1988; Brint, 1994). The identifying feature of the scientists who compose epistemic communities is that they act as experts in the policy world. In a context of transnational policy-making, this means that they are infiltrated in a variety of organisations and arenas that contribute to make policies. Epistemic communities are dispersed, and 'do not exist as separate collectives but are embedded in other types of collectives, and especially in formal organizations' (Mayntz, 2010).

These three factors combined means that politicians and bureaucrats, deep in the world of action, may turn to experts and their ideas at particular junctures. The influence of the latter depends on when, where and how this transmission between the two worlds is done (Haas, 1992). The influence of experts only materialises when there is a connection between these two worlds. The usual connection that is considered in studies of epistemic communities is that of scientific or expert advice to politicians. This channel allows experts to transfer knowledge beyond the frontiers of science into policy-making. The activation of such a channel presents many of the problems of the analyst: by which process are these channels activated, who controls these channels, what happens with the elements that feed into policy, and so on? Bureaucrats and politicians are the gatekeepers (Haas, 1992; Marier, 2008). Expert advice, to use Churchill's famous word, is on tap not on top.

The notion of epistemic community is continuous with that of network as Haas shows by his definition ('network of professionals'). The network notion was mobilised in other literatures to emphasise similar dimensions of policy coordination through ideas, expertise and information exchange. Extending the study of international regimes and moving away from the assumption that states are unitary actors, Slaughter (2004) as an international legal scholar sets the focus on the mobility and transnational cooperation of government officials and other state personnel such as regulators, judges and legislators who compose transgovernmental networks. She defines a governmental network as 'a pattern of regulatory and purposive relations among like government units working across the borders that divide countries from one another and that demarcate the "domestic" from the "international" sphere' (Slaughter, 2004: 2). Her theory to

explain the proliferation of such networks is a functionalist one: governments need to extend their reach beyond national borders in a world where
most of the issues that governments have to deal with unfold on a global
scale. Without actually using the term 'risk', she argues that threats and
disasters are particularly significant drivers of the transnationalisation of
governments: 'Networked threats require a networked response [. . .] In a
world of global markets, global travel, and global information networks,
of weapons of mass destruction and looming environmental disasters
of global magnitude, governments must have global reach' (Slaughter,
2004: 4). In such transnational networks, state personnel appear less in
their domestic identity but more as international experts. The need to
gather expertise is also an integral element of the transnationalisation of
government.

Scholars of comparative policy analysis also employ the notion of
networks to contrast different types of national policy structures.[1] It
also appears useful to address the Europeanisation and globalisation
of policy-making. This literature highlighted the role of transnational
policy networks (Bennett and Howlett, 1992) or global public policy networks (Reinicke, 1998; Stone, 2004) in these new modes of policy change.
Authors in this stream of work emphasise the importance of a kind of
soft power residing in information exchange, circulation of ideas or policy
learning. All of these phenomena are at the root of the convergence of
actors' beliefs and preferences, and of policy change more generally.

The notion of government or policy network is continuous with other
notions that emphasise the sharing of cognitive and normative elements among policy-making actors (Skogstad, 2005), such as epistemic
communities and advocacy coalitions (Sabatier, 1988).[2] There is no
question of science policy here, but this network approach nonetheless
provides explanatory elements for the understanding of scientific expertise
and transnational regulation. First, networks mark the specialisation of
policy-making and regulation. They unfold strictly within the boundary
of a domain or issue area, and comprise specialists, often experts on the
particular issue. Second, one of the founding arguments for the use of the
notion of network is the need to cross the boundaries of the state and government and recognise the participation of professional and private actors
in policy-making.[3] And indeed most publications quoted above have references to experts and expertise as integral actors of transnational networks.
A further illustration of the value of the notion of network is to reflect
that there are separate networks formed around activities of knowledge
production, evaluation and information exchange, which cumulatively
have an impact on policy-making or rule-setting. As a counterpart to the
focus on governmental actors, other notions indeed have been conceived

to assert the role of 'non-state actors' in transnational governance (Keck and Sikkink, 1998; Risse-Kappen, 1995), and of knowledge agencies in particular. This includes the 'embedded knowledge networks' (Sinclair, 2000) and international or global knowledge networks (Parmar, 2002; Stone, 2002, 2005). While embedded knowledge networks seem quite specific to the world of financial markets and regulation (they cover credit rating agencies and the world of financial information by which integrated judgements of creditworthiness circulate), the latter have been usefully defined in a generic way:

> Knowledge networks incorporate professional associations, academic research groups and scientific communities that organize around a special subject matter or issue. Individual or institutional inclusion in such networks is based upon professional and official recognition of expert authority as well as more subtle and informal processes of validating scholarly and scientific credibility. The primary motivation of such networks is to advance, share and spread knowledge. While many of these arrangements are engaged in the disinterested pursuit of knowledge, they can also be policy focused. (Stone, 2002: 2)

These networks span research and policy. For Parmar (2002), they stimulate and coordinate the production of new knowledge and ideas about policy issues (such as in this case, economic development). Networks of think tanks as organisations populated by thinkers and intellectuals with an explicit agenda of diffusion of policy ideas, offer an example of the often strong ideological orientation of these networks. Stone (2005) offers interesting insights about knowledge networks that would easily apply to the standardisation of regulatory activities. They systematise knowledge and impose that which is codified, technocratic, secular and westernised. They manage to do this through their social authority and credentials, and with their political resources. They have close relationships with policy-makers, and combine 'epistemic, discursive and ideological practices'. This results in particular cognitive dynamics of construction of public problems, which in turn, serves to centre the policy-making process on admitted stakeholders and to reinforce the strength of the knowledge and policy network.

WHAT SORT OF SOCIO-INTELLECTUAL FORMATIONS ALLOW STANDARDISATION?

What do these concepts of epistemic community and knowledge network hold for the purpose of this book? These different labels for transnational collective actors are specifically useful in emphasising the density of interactions and relationships that now exists among scientific experts, including transnationally, as well as between the latter and regulators or policy-makers. These notions open a possible claim, though not always a demonstration, that regulation is being constructed transnationally. On this particular scale, the influence of scientific experts on regulation and policies seems greater than in domestic politics precisely because scientists and professionals have long been active at this level – probably for a longer time and more legitimately than governmental actors (Wuthnow, 1979). They also give substance to the idea that regulation is made through processes of exchange of ideas, information and learning, in the name and under the ruling of those who are believed to be experts in the area (Majone, 1989).

There is one major drawback, however, to using them. The causal link between the organising of experts as a community or network, and the capacity to transfer research findings, information, ideas and standards into policy remains unclear.[4] Little empirical work is done to make visible the mode of interaction and type of relationship that justify calling the collective a community or network, and that explain their ability to produce shared and standard knowledge (Djelic, 2004). Such demonstration would however be welcome, given that epistemic communities are quite extraordinary things, of quite rare occurrence. It may easily be admitted that knowledge is shared and is a ground for strong communal ties. But communities of knowledge production, of science in action as it were, are not the most frequent cases. And third, when this knowledge has policy and political aspects to it, it becomes even more unlikely that it is generated by communities.

In reality, Haas' paper does show how distinctive or exceptional an epistemic community is as a type of socio-epistemic formation. It is a network of professionals who share a set of causal and moral beliefs, and are engaged in a collective policy enterprise. On the one hand, sharing these beliefs differentiates such a network from an interest group. On the other hand, pursuing a policy enterprise distinguishes them from scientific disciplines and professions (along with their moral beliefs). But in addition, pursuing such enterprises does not equate epistemic communities with interest groups though, because there is a limit for members of an epistemic community regarding how far they will follow this policy interest. They will not go against their beliefs in the pursuit of this policy agenda. Thus, the combination of these two features makes the epistemic community a new *sui generis* community.

This unique characterisation is the result of an original approach by Haas, who found inspiration in the sociology of scientific communities but was interested in explaining the existence of a set of norms and rules around which expectations converge at the international level (Keohane and Nye, 1977; Adler and Haas, 1992). All sorts of professionals, not just natural scientists, may form an epistemic community and gather influence over policies. This is possible as long as they are professionals; that is, that their specialised knowledge is socially valued and authoritative. But the original emphasis on beliefs comes from the study of scientific communities, one of Haas' sources of inspiration for the concept of epistemic community (Haas, 1989). Haas indeed quotes Ben-David, a sociologist of science, saying that epistemic communities with their tightly shared beliefs demonstrate 'one of the interesting instances where a group of people is held together by a common purpose and shared norms without the need of reinforcement by familial, ecological, or political ties' (Ben-David, 1984; cited in Haas, 1992: 20). This quote and the whole epistemic community theme resonates with the identification by Merton of the norms of science, and notably the perception that all findings of science are commonly owned by all scientists (Merton and Storer, 1973). The reference to Ben-David inscribes the theme of epistemic communities in an older tradition in the sociology of science that looks at 'thought collectives' (Fleck and Trenn, 1981) – as collective repositories of credentialised knowledge and proponents of particular paradigms – or at 'scientific communities' (Kuhn, 1970), characterised by the sharing of a set of values, of problems and expertise, which is also marked by intense internal interactions.

But first identifying shared beliefs remains a methodological challenge. Keohane (one of the initiators of the study of international regimes, a root of research on epistemic communities) expressed doubts about this very possibility:

> It is enormously difficult, indeed ultimately impossible, to determine 'principled and shared understandings'. To what extent principled, to what extent shared? How are we to enter into the minds of human beings to determine this? And which human beings will count? Even if we could devise a way to assess convergent expectations intersubjectively, what standard of convergence would we require to determine that a regime existed? (Keohane, 1993; quoted by Hasenclever et al., 1996: 182).

Authors who write about epistemic communities are not immune to this criticism. But they circumvent this difficulty by looking exclusively at contexts where a certain amount of sharing of information, knowledge and norms already exists by definition, since they are interested in explaining the existence of international regulatory regimes. From that point, they trace the relations between experts involved and tease out the elements of knowledge and beliefs that they share. The reverse heuristic – following one or several scientists through their career and research to see whether they converge towards others to form a community as they get involved in policy-making – has not been employed.

Second, there can be no such thing as stable and influential expert communities. This is because the disputed, unstable or transient nature of scientific knowledge removes this possibility – even more so in contexts of risk and uncertainty, in which epistemic communities are most influential according to Haas (1992). Sociology of science, from the 1970s onwards, building on its distinctive interest in the production of knowledge and the social determinacies that bear on it, has further pushed this critique of the sociological motif of the community. The production of knowledge is not enclosed within the boundaries of well-delimited groups; it does not respect social demarcations. It involves the exchange of tacit information that is not traceable through sociometric indexes (Collins, 1974), informal contacts (Jacobs, 1987) or institutional migrations (Hoch, 1987). It cannot be explained by apparent social relationships and memberships (Collins, 1974). Different concepts were then produced to account for the transient and unstable nature of knowledge: core-sets (Collins, 1974 and 1981b), research networks (Mulkay et al., 1975) or transepistemic arenas (Knorr-Cetina, 1982). The latter concept of 'arena' definitively invalidates that of community. It makes clear that the notion of community can only be imposed from the outside by the analyst on a set of scientists, but that these scientists would not perceive themselves as being part of a community. To be sure, being involved in standard-setting activities is not a shared characteristic of a particular community. Knorr-Cetina swaps the notion of membership or belonging to a community for that of common knowledge – such as the 'points of reference' (Gusfield, 1975), standards of proof, criteria of scientificity, and so on. This knowledge constitutes a material and symbolic resource that scientists can mobilise.

This approach of scientific collectives is in direct opposition to attempts to identify communities. It is interested instead in the institutional territories occupied thanks to the stabilisation of bodies of knowledge. In contrast, a more constructivist sociology of science focuses on open and flexible networks that extend across disciplinary boundaries and the negotiation of these boundaries (Merz, 2006). Its scepticism extends to the hypothesis of expert communities, for at least three reasons. The first is that scientific expertise is not a particular type of knowledge or professional identity. It is a situation in which knowledge claims are criticised and tested according to criteria of validity that differ from those in which they were first produced. Expertise has no natural condition (Edmond, 2004). It it best understood contextually, as an encounter between a problematic situation and a knowledge, selected and constructed through specific criteria of relevance and facticity in that context (Fritsch, 1985;

Wynne, 1989). Second, the ability to take on a multiplicity of commissions and activities outside academia is not solely the preserve of those scientists who act as experts in policy circles. Any scientist, in fact, is led to such situations along his or her career. No scientist is purely an academic or researcher: all scientists have a social role as an expert, because the stabilisation and authority of knowledge claims necessitate acting beyond the sites in which this knowledge is produced – laboratories and university departments. Summarising the argument of Karin Knorr-Cetina, Struan Jacobs (1987: 274) writes that:

> a scientist's research decisions and selections are influenced by people beyond his/her laboratory and beyond any given specialist area. Connections are formed with scientists in diverse subjects, with government officials, university administrators, lay representatives of private corporations, any of whom may affect the form and content of research projects.

Third, the presence of strong uncertainties around the causes of the problem being treated undermines the chance for scientific knowledge to gain ascendance and authority. On the contrary, a situation of uncertainty tends to turn into scientific dispute with challenges to the authority and credibility of knowledge claims. Uncertainty increases competition between proponents of different scientific interpretations. It generates disagreements, suspicions concerning the motivations of scientists, and interrogations concerning their assumptions and methods (Nelkin, 1984). Communities are thus unlikely to appear in a climate of uncertainty and scientific dispute. They may only emerge once the controversy is closed:

> In technically grounded controversies in the policy domain, the central question most often is what is going to count as relevant knowledge in the first place; who possesses the right sorts of knowledge is secondary. 'Core-sets' may sometimes form (in decades of research I have observed such groups forming, for example, around 'dioxin science' and cancer-risk assessment), but these bodies of mutual accreditation and acknowledgment are very often the result of protracted controversy. Just as in sociology of science, 'facts' are depicted as the results (not the causes) of closure of controversy, so 'core-sets' in public domain technical controversies are the results, not the causes, of controversies settling into normalized patterns of debate. Importantly, it is when debates finally lose their public significance that one often sees these exclusive knowledge clubs emerging; we have a recognized cadre of 'dioxin scientists' today for just the reason that no one cares much any more about the proliferation of dioxins in the environment. That battle has been fought, and largely won, by those inclined to precaution with environmental toxins. (Jasanoff, 2003: 395)

In this perspective, scientific communities or sets of scientists can at best emerge outside episodes of controversy when scientific knowledge gains acceptance and criticism declines. In other circumstances, boundaries tend to become porous. Scientists involved in controversies have difficulty asserting the scientific and non-politicised nature of their knowledge. The science–politics boundary gets eroded, making it more difficult for scientists to uncontroversially appear as experts. In other words, the research on epistemic communities does not seem to overcome criticism that comes from constructivist sociology of science. At best, epistemic communities seem to be a rather exceptional occurrence. Other concepts do not seem to do any better, or have not led to more sophisticated methodological measurement of circulation and informal associations. Nowhere in the literature on policy or knowledge networks can a proper social-structural analysis of positions and interactions between experts be found, as understood in the pure sociological sense of network (Burt, 1995).[5] Overall, there is a tendency to grant all sorts of very important properties to these collective actors without much of an explanation for how they achieve so much.

More succinctly, the second reason for not adopting these approaches is the focus on already made and validated knowledge, rather than on 'science in action' (Latour, 1987). Scientific experts are said to 'carry' (other terms include 'transfer', 'diffuse', 'spread') an already-constituted knowledge, which is then carried over into policy without transformation.[6] It is produced on one side and used on another. Communities and networks are vectors, not sites of knowledge production. In spite of the allusion of Stone (2005) that knowledge and policy are mutually constitutive, there is little indication of how the epistemic action of experts is constitutive of policy intervention, and of why and how expert networks or communities would be propitious sites for the production of regulatory knowledge. The overall approach to expert scientific knowledge that underpins these literatures is also quite positivistic. Expertise and politics are easily distinguished, politics being the realm of value conflicts and power plays, science that of solving uncertainties.[7] However, expertise may also be defined as the encounter between specialised knowledge and a problematic situation (Fritsch, 1985; Edmond, 2004), an inherently political process. The reference to an authoritative expertise prevents us from taking into account the many meanings of expertise (Nunn, 2008)[8] and the many ways in which this encounter takes place. The term 'expert' or 'scientific expert' may apply to people with very different types of knowledge production and epistemologies, as well as practices of engagement with political actors (Waterton, 2005). Overall, there is a tendency with such notions of epistemic community and knowledge network to consider 'scientific expert' to be a set position beside politics, but connected to policy-making by close relationships with policy-makers and the manipulation of such resources as information and technical knowledge. But it leaves out the possibility that expertise is constructed from within politics.

INVISIBLE COLLEGES OF STANDARDISATION

It may be helpful to start from an altogether different premise to understand the constitutive nature of scientific expertise for regulatory action. Being a scientific expert in risk regulation does not mean occupying a set position in the regulatory space. It is a pattern of inscription in the regulatory world that is characteristed by a form of mobility. There are several important implications of this circulation, which the concept of invisible college aptly incorporates. The concept comes from the history of science and scientometrics, but was initially more of a metaphor than anything else. The expression is the term chosen by the pioneer English scientist Robert Boyle to speak of an unlikely grouping of Anglo-Irish men of science around him and Benjamin Worsley, who all shared the same conception of experimental natural philosophy as well as philanthropic, universalist and utopian aspirations (Webster, 1974). By 'college', Boyle was most likely making reference to the notion, already in use at the time in the universities of Oxford, Cambridge or Sorbonne, of a society of scholars incorporated within a university. There is considerable debate about the choice of the adjective 'invisible'. The historiography seems to have come to an agreement that the term refers to the informality of Boyle's group, that was not incorporated within a university and was also independent from the then developing scientific societies such as the Royal Society of London for the Improvement of Natural Knowledge founded in 1660, just a few years after Boyle made the first reference to an invisible college in his correspondence.[9] The invisibility of this college otherwise refers to the fact that the group were maintained mainly through a dense literary correspondence but only sparse meetings, as in the case of Boyle and of his associates who were geographically dispersed across Europe. In this respect, the notion of invisible college refers to 'a collective body' or 'assemblage' rather than 'an organized society of persons' (*Oxford English Dictionary*).

The term was taken up by Price (1963) to study groups of scientific elites and later by Crane (1969, 1972), to form a well-established research tradition in information sciences and bibliographic research (Price, 1971; Zaltman, 1974; Noma, 1984). With this term, Price was interested in accounting for the observed concentration within limited groups of the publication of most-cited academic articles (in the context of the exponential number of academic articles in any growing research field):

If there are 100 authors, and the most prolific has a score of 100 papers, half of all the papers will have been written by the 10 highest scorers, and the other half by those with fewer than 10 papers each. In fact, in this ideal case, a full quarter of the papers have been written by the top two men, and another quarter by those who publish only one or two items. (Price, 1963: 46)

This intuition about science's de facto elitism was later tested by Diana Crane, who empirically measured social ties between scientists to try to find out whether there were special relationships between the scientists belonging to these restricted groups. Through questionnaires, she studied the social relationships between members of a research area. She demonstrated that a research area is constituted of two sorts of groups. The first kind are groups of collaborators who have a solidarity function in socialising new scholars, organising research, providing a sense of belonging, and so on. It is within these groups that experiments are done, publications written and reviewed among other things. The other is the invisible college. She demonstrated that some groups lead and orient scientific production, while others follow. She gave evidence to show that rather narrow groups of scientists have a strong influence over the course of research in any given area. Crane's other major insight is that the invisible colleges of Price influenced the trajectory of a research specialty (Crane, 1972). Her research established that the logistic curve that any research specialty follows in terms of number of publications (a launch phase with a slowly increasing number of publications, followed by a period of intense activity and exponential number of publications, then declining activity and number of publications) is explained by the underlying social structure of research specialties. In a sense, Crane solved the difficulty of measuring transient, informal and implicit knowledge exchanges by assuming that publishing around a set of concepts and participating in the scientific production of a field could be used as a proxy for the sharing of beliefs. She applied these findings to areas of international policies, noting that transnational invisible colleges were particularly effective at setting up international working groups (Crane, 1981) and showing that the hypothesis that such colleges form transnationally is a plausible one. She argues that invisible colleges activate the consultative relationships from which international scientific and professional associations benefit, almost statutorily. Through close and repeated interaction with the international governmental and intergovernmental organisations active in their field, they contribute to the integration of highly diversified and fragmented programmes.

The concept of invisible college is not specifically designed to apply to matters of science in policy. Furthermore, the work of its early proponents such as Derek Price and Diana Crane falls under the criticism of the 'community' motif of later sociologists of science, that it over-estimates the degree to which membership in particular social groups can explain knowledge production (Collins, 1974). The work of Price and Crane is rooted in scientometrics and makes a great case to study scientific production quantitatively. It is the source of research on scientific specialties (Edge et al., 1976), which have been strongly criticised by the other sociology of science that emerged in the second half of the 1970s and that preferred to observe with forensic precision life in laboratories and other sites of knowledge production (Chubin, 1985). One of the criticisms raised about Crane is that she does not account for informal relationships. Knowledge production is based on sets of open, tacit and transient relationships that traverse social boundaries. The notion of invisible college does badly at taking into account these tacit and highly informal relationships. Were Price and Crane to consider informal relationships between members of invisible colleges as distant geographical affiliates exchanging experimental and informal knowledge in the form of draft papers or pre-prints for example, they would still be looking at quite 'organised' evidence of these relationships producing co-authored papers (Lievrouw, 1989; Zuccala, 2006). There are nonetheless three reasons that indicate that the concept may have some value in helping to analyse the relation between scientific evaluation of adverse events and the constitution of standards of regulation.

The first is methodological. The sociographical work performed by Crane cannot be discounted in a context in which, after decades of enquiry into the complex matter of science and expertise, we are still left wondering who experts really are, what their careers are like, what explains their involvement in regulation or what comprises their collective identities (Cozzens and Woodhouse, 1995; Joly, 2007). The constructivist sociology of science with its distinctive focus on controversies and other contexts of production and diffusion of knowledge claims, has clearly overlooked these questions. Authors such as Jasanoff while not completely eliminating these aspects from their research, certainly considered them marginal to their analysis. Because of this, and because of a lack of investigation into the interests, identities and social structures of expertise, it is not known why certain scientists seek to be recognised as regulatory scientists or experts, while others are content with being in academia proper. Studies on professional and disciplinary identities ceased in the mid-1980s (Robbins and Johnston, 1975; Lemaine et al., 1976; Cambrosio and Keating, 1983; Wray, 2005) and have not been carried over to the study of scientists in regulation. Now, there is a lot to gain in combining quantitative and qualitative methods as performed by Crane with the understanding of scientists' policy expertise and influence, particularly when scientists have such complex professional trajectories and identities. The concept also denotes the elitist or hierarchised nature of scientific fields, in which all scientists are not equally influential and authoritative. Arguably, this elitism is a feature that scientific research fields have in common with regulatory science fields. The capacity to take part in for instance the regulation of biotechnological foods and to influence it by forging its standards is not equally distributed among all the scientists who do laboratory research on these kinds of foods. The cumulation of academic credentials and of missions or roles in regulation is the characteristic of a scientific elite, and it is essential to its ability to exert a 'mediating role' (Mulkay, 1976). It seems important, in these matters of risk, to be able to account for the evaluative posture of scientists interested in risks and their assessment, by which what they assess is always necessarily an adverse event, its likelihood, but also the regulatory conditions that make it more or less likely. This evaluative posture is characteristic of the elite segments of disciplines and professions, to which the scientists followed in this book belong.

The second reason is the reference to the multiple contexts in which scientists simultaneously act when they develop an interest in public matters such as risk regulation. Scientists interested in evaluating health risks – some at least – are characteristically very mobile and polyvalent. They do not stay prisoner of the labs and lecture halls in which they often started their career. Their authority and reputation create opportunities to enlarge their portfolio of professional activities. As Jasanoff puts it, 'Protected by the umbrella of expertise, advisory committee members in fact are free to serve in widely divergent professional capacities: technical consultants, educators, peer reviewers, policy advocates, mediators, and even as judges' (Jasanoff, 1990a: 237). This ability to wear different hats in different contexts, successively (at different stages of a career) or simultaneously using the credibility and authority they enjoy as holders of specialised knowledge, defines what a scientific expert is. They are indeed not just researchers, academics, professionals or governmental advisers but 'multi-professionals' (Shinn, 2005). This, anachronically speaking, is one of the meanings of invisible college since Boyle referred by this term to such polyvalent people. Circulation matters in so far as it explains a willingness and possibly also an ability of certain scientists to accumulate practical experience of adverse events and of their control, to produce a contextual knowledge, such as the typifications Scott speaks about (Scott, 1998; see Chapter 1).

Analytically, circulation is across a two-dimensional space. Horizontally, circulation is between the different actors and activities of control. As is well known and is the reason why there are now 'conflicts of interest' and transparency rules for scientific advisers, members of scientific committees are often also professionals who study the products they evaluate as public risk assessors (or who prescribe these products in the case of physicians). Even where scientists do not participate in product development they may well be conducting experiments on novel substances that companies may at the same time be trying to incorporate into foodstuffs. As specialists on a product or kind of adverse event, these scientists may also, and in practice do quite often, participate in public debates and get interrogated by journalists. Being themselves researchers or ex-researchers, they also continuously exchange with laboratory scientists. In general, this means that they circulate between teaching and research, consulting activities, membership of policy and standard-setting committees at national or international level. They participate in regulatory activities in various guises, not only through product approval and risk assessment but also through monitoring and surveillance of products in the market, evaluation of companies' compliance, as well as more high-level and strategic development of governance schemes (Doern and Reed, 2001). The vertical dimension of this circulation is between activities of control on the one hand and activities of standard-setting and their organisations on the other hand, such as the International Conference on Harmonisation of Technical Requirements for Registration of Pharmaceuticals for Human Use (ICH) or the Codex Alimentarius (Codex) (that for reasons of legitimacy need and indeed seek closer association with experts; Brunsson and Jacobsson, 2000; Tamm-Hallström, 2004) or international policy-making organisations such as the European Commission or the World Health Organization. In sum, scientists circulate across regulation and do so transnationally.

The third and final reason for using this concept is that it aptly captures a pattern of mobilisation. This circulation brings scientists of one domain back to the same places, in such a way that over time they get to meet regularly and form relationships. Scientists are not 'affiliated' to or members of a college. They form part of a loose collective. The ties between them are typically weak, intermittent and irregular – this is after all what the adjective invisible, in 'invisible college' refers to. Conversely, the term college is here to emphasise the fact that contingently, segment by segment, the circulation of the same scientists in the same places delimits a circle of experts who may be gathered together in certain circumstances. Scientists who are in public administration or scientific committees, for instance, do not suddenly abstract themselves from older relationships with colleagues employed by the industry or in academia. Even though they do not meet all together at one place and time, the traces of these links and encounters remain. Common knowledge is not necessarily stored in one place, constituted as a paradigm or disciplinary corpus. But it still circulates through scientific publications, expert or policy reports in professional literature or draft standards that scientists exchange among themselves. Price and later Crane did highlight these features of circulation, if only in passing. Price used the expression of invisible college to denote that the most productive and most cited scientists in a field are geographically distant affiliates, who exchange information and 'pre-prints' (Price 1963: 84). The organisation of science into colleges 'is not perfect': all important people are not always in the same place at the same time, some who would qualify to be in the group indeed never come. These people also 'meet piece-meal'. A college is above all a 'commuting circuit', 'so that over an interval of a few years everybody who is anybody has worked with everybody else in the same category' (Price, 1963: 84–85). Experts formally meet in interstitial organisations. Iterative encounters result in the setting up of standard-setting platforms that are often interstitial such as expert committees, expert consultations, industry-funded scientific foundations, think tanks or professional associations. Interstitial organisations, different in this regard from 'boundary-organizations' (Guston, 2001) are characterised by deliberately faint or absent affiliation with vested interests to better appear as a neutral space for the construction of consensus. Experts in interstitial organisations are not affiliated or accountable to a formal organisation, or distance themselves from these affiliations. Being interstitial they are not offending any other vested interest. Invisible colleges do not displace existing memberships, but provide a subsidiary and inoffensive space where they can test and exchange ideas in a way that they would not under a stricter regime of affiliation.

CONCLUSION

The notion of invisible college is therefore useful in underlining the elitism of science and scientific expertise in regulation, and in studying experts empirically. It emphasises the impact resulting from the elitism (in the activity and science of safety evaluation), circulation (practical and generic experience of adverse events) and interstitiality (regulatory influence) of small segments of scientists who seem to form part of a common circuit. Scientists circulate and meet among themselves, even in a piece-meal fashion, to gradually accumulate and standardise experience of adverse events and modes of control. In contrast with social authority and cultural legitimacy, these are the factors inspired by history of science and scientometrics.

The following chapters will serve to illustrate how these colleges can observe and channel experience of adverse events and of different types of intervention to forge regulatory concepts. Invisible colleges connect myriad changes occurring in different places. They connect the experience of different companies and regulatory agencies with similar types of adverse events and work to compare these experiences. Concepts crystallise through a web of scientific publications, reports, opinions of scientific advisory committees, academic and policy conferences, public debates and think-tank publications. As will be shown later, business service companies also play a critical role, such as through regulatory bodies and risk-management consultancies, contract research organisations, certification companies, all of which monitor and anticipate regulatory changes to define which service manufacturing companies may need in the future. Devices, procedures and equipment start being elaborated that will implement the process, such as pharmaceutical risk-management training, HACCP manuals and guidelines, software for the management of safety databases, and so on. Invisible colleges are the embodiment of these complex and vast circuits of standardisation of safety controls.

The intention here is not to argue either that the invisible college is *the* form of transnational collectives of experts, that it can be found everywhere we look – a tendency with the invasive notion of epistemic community and even more so of network – or again that expert scientists are exclusively affiliated to these (Schott, 1993). Colleges do not form at any time or anywhere to codify aspects of risk regulation but are on the contrary quite contingent. Invisible colleges should not be confused with small and exclusive elite clubs (Van Apeldoorn 2000; Graz, 2003). The notion of invisible college serves here to show that scientists and their evaluative knowledge circulate, and that they link together more formal standard-setting arenas. It is used to speak about a particular pattern of association and knowledge production that mixes institutional precariousness and epistemic power. Besides being reflective of the properties of transnational communities – notably their contingency, interplay with formal organising, their embeddedness in more local or national sites (Djelic and Quack, 2010; Mayntz, 2010) – the notion helps to make sense of the paradox by which the diversity of scientists' individual trajectories and their range of knowledge enhances rather than jeopardises their ability to mediate the actors of a regulatory domain. It suggests that what matters in the emergence of regulatory standards is not only that scientists sharing the same ideas speak to policy-makers and do so simultaneously but that they circulate between actors of the domain and transform experience of control over adverse events into common concepts and standards.

NOTES

1. With the globalisation of policy-making, the academic fields of public policy and international relations have been seen to converge (Risse-Kappen, 1996; Coleman and Perle, 1999; Petiteville and Smith, 2006), and the notion of networks is one of the elements that they have come to share.
2. This reflects the definitions originated by Marsh and Rhodes (1992), for whom policy communities are integrated, stable and exclusive policy networks. Issue networks, by contrast, comprise loosely connected, multiple, and often conflict-ridden members.
3. The notion of policy network comes from the field of public policy, building on Heclo's notion of issue networks (Heclo, 1978), British tropism for governance (Marsh and Rhodes, 1992; Rhodes, 2000, 2007), and a German tradition of analysis of interorganisational relations in the public sector (Mayntz and Scharpf, 1975). The notion has been useful in breaking down the illusion of state unity and to bring more complexity and politics into the production of policies and legislation (Atkinson and Coleman, 1989; Marin and Mayntz, 1991). There is neither an agreed definition of what a network is (Börzel, 1998), nor even an agreement on whether policy network is a concept or a metaphor. There does not seem to be an overarching tradition of analysis in this literature either (Thatcher, 1998).
4. See Dowding (1995) for a more specific critique of the explanatory power of policy networks.
5. See the study of the transnational 'governance network' of central bankers by Marcussen (2006) for a counter-example.
6. Hence a close association of the notions of transnational communities or networks with the literature on policy diffusion (Busch and Jorgens, 2005). This literature however does not sufficiently acknowledge the fact that a policy or instrument, as it travels, gets transformed (see Dratwa, 2004; Djelic, 2001; Fourcade, 2006; and James and Lodge, 2003 for different expressions of this same note of caution).
7. In the case of epistemic communities, this boundary between knowledge and policy is often expressed in terms of policy-making stages (Jones, 1970). Scientific experts may well be active within administrations but their role is one of scientific advice. They act upstream, and are rarely involved beyond the point of decision-making. In certain cases, they are shown to influence the design, diffusion and defence of policy solutions (Hasenclaver et al., 1996). But they are absent from policy implementation, i.e. from the actual interventions of policy and all that they generate. In contrast, the practical experience and involvement of scientists is here shown to be key in the capacity to conceptualise regulatory intervention.
8. It is quite striking that expertise is a recurrent theme in the literature on transnational or global governance. It is often presented as a power resource. This is taken for granted rather than problematised though. Definitions of expertise are seldom found and book indexes rarely include this item.
9. As a consequence and mistakenly (Webster, 1974), the invisible college is frequently cited as the prefiguration of the Royal Society (Price, 1971; Price and Beaver, 1996).

3. From qualifying products to imputing adverse events: a short history of risk regulation

Evaluating risks is a primary source of regulatory standards. In the area of food and pharmaceuticals safety, evaluation of risk has become a science. Disciplines such as clinical pharmacology, regulatory toxicology or microbiological risk assessment are founded, more or less explicitly, on the principle of classifying or ranking products according to their safety, rather than simply experimenting, observing or theorising the latter. These evaluative sciences are 'trans-sciences' or 'post-normal' (see Chapter 1) in so far as they are implicitly normative: they are the source of safety criteria and specification against which this act of classification or ranking can be made.

Evaluative sciences have evolved from a concern with the qualification of products and their safety to one in which they try to anticipate and attribute adverse events in spite of the uncertainty as to when and where they might occur. Qualification is a safety approach that covers the tools and criteria to evaluate the qualities of a product at a given point in time. It covers clinical trials, animal toxicity-testing models, microbiological models, all sorts of experimental techniques by which the hazardousness of a product may be predicted. A programme of qualification aims to shape markets and consumer behaviour by ascertaining through controlled and reliable knowledge, the qualities of a product – positive ('efficacy') and also more negative ones ('side-effects'). It is organised by the standards it produces, such as testing guidelines to verify and project claims about these qualities to the outside world. The programme of imputation differs in several important ways. The origin of the criterion of safety to evaluate is empirical; it is constituted by the many signals that arise from the use of products. It is indeed necessary to detect and gather these signals because of the assumption, deriving from experience of disasters and other unforeseen cases, that uncertainty is structural. Imputation acknowledges the open-endedness and causal complexity of risks. It is about building as comprehensive as possible an image of the adverse events that occur out there and working towards their causes. In line with epidemiological thinking, it aims to typify adverse events and construct an image of the reality of risks through regularities, as opposed to a programme of qualification that is oriented towards the setting of individual product specification. Imputation covers the setting up of other analytical devices, systems of observation and information gathering, methodologies by which signals of adverse events are monitored and captured, in full recognition of the uncertainty surrounding when, where and with what severity they will happen. Imputation also aims to trace the causes and responsibilities of these uncertain adverse events. This includes regulatory responsibilities, or the limitations and failures of activities of control to avert adverse events. In this, it is the research programme on risk par excellence (Luhmann, 1993). Evaluative sciences always evaluate adverse events and regulation at one and the same time.

Qualification and imputation constitute scientific research programmes in their own right, which can be traced through academic publications and scientists' discourse. By showing how European laws eventually reflected these two programmes, and specifically a shift towards imputation that took place gradually in the second half of the twentieth century, this chapter reinforces the hypothesis (which subsequent chapters will test and explain by studying emerging standards of imputation) that science is the source of standards of risk regulation. The chapter covers a long period and many standards. The breadth of coverage will possibly appear to be at the detriment of depth. But it is necessary to draw the landscape of food and pharmaceutical regulation. First, I describe the regimes of regulation of food and pharmaceuticals in the EU, tracing them back to the major disasters that in great part explain their adoption. This is aimed to create a contrast with the demonstration, in the subsequent parts of the chapter, that these laws derive from the evaluative work of scientists in the intervals of major disasters and on longer series of adverse events. A scientometric study helps me to capture the scientific programmes that underpin evaluation and extract particular concepts and ideas that, as I finally show, have filtered into European food and pharmaceutical laws.

TESTING PRODUCTS/MONITORING RISKS

The consumption of contaminated foodstuffs and unsafe medicines regularly causes disastrous consequences, both in terms of the health of people and number of fatalities as well as in political symbolic terms. While the production of meta-assessments of risks and lack of safety is fraught with methodological difficulties, it may nonetheless be said that both constitute a very serious public health problem. A highly quoted study of adverse drug reactions (ADRs) in hospitalised patients by Lazarou et al. showed that the total incidence of both categories of serious ADRs was 6.7%, with an overall fatality rate of 0.32% (1998). A recent Swedish study has also implicated ADRs as the seventh most common cause of death (Wester et al., 2008). In a study of almost 19,000 admissions, other authors were able to show that 6.5% of patient admissions to two National Health Service (NHS) hospitals in the UK were related to an ADR (Pirmohamed et al., 2004). Food contaminations and poisoning are even more varied (they may involve a microorganism, a chemical substance, naturally occurring toxins or other agents such as the prion that causes Bovine Spongiform Encephalopathy (BSE)) and are often not reported. But the WHO for instance (WHO, 2007) argue that in 2005 alone 1.8 million people died from diarrhoeal diseases, many of which could be attributed to contaminated food and water. In developed countries, an estimated 30% of the population each year suffers from foodborne illness. Certain foodborne epidemics can cause waves of fatalities, in which case they often lead to strong policy reactions.

In any case, food and pharmaceutical safety are areas with great potential for health disasters and policy reforms. The main controls for food and pharmaceutical safety in the EU derive from original crises such as that of thalidomide for medicine and BSE, for food safety (Krapohl, 2008). Both regimes have in different ways and at different times followed on from these more or less tragic episodes that construct policy failures, trigger a search for responsibilities and facilitate policy reforms (Kingdon, 1984; Bovens and t'Hart, 1996). These events qualify as political crises in that they legitimise new policy enterprises and the passage of new legislation by creating 'dissatisfaction' with regulatory policy – a climate from which political entrepreneurs can strive to provide the initial impetus for qualitative shifts (Harris and Milkis, 1996). Crises create a rotation of elites, with new groups of policy-makers entering with a stock of new solutions to respond to failures (Keeler, 1993). I propose a little detour via US drug regulation to illustrate this pattern of regulatory change and as a point of departure for the historical trends that are traced here, that is the emergence of standards and tools to establish causalities and impute even highly uncertain events.[1]

One of the founding laws of modern drug regulation – and modern regulation in general – is the Federal Food, Drug and Cosmetic Act of 1938. This Act decisively gave the power to the US Food and Drug Administration (FDA) to reject the marketing of a drug ex ante. For Carpenter (2010), this Act is a 'commemoration in law' of the tragedy of the 107 deaths caused in 1937 by the consumption of Elixir Sulfanilamide, a preparation containing a toxic solvent and wrongly labelled as an 'elixir'.

This disaster finally helped to pass a bill that had been pushed by the Agriculture department and Senator Royal Copeland to strengthen the powers of the FDA, but had previously failed to be passed several times (Quirk, 1980; Temin, 1980 and 1985; Marks, 1997; Carpenter, 2010). The health and policy tragedy of the Elixir finally propelled the bill onto the Senate agenda for it to become law. The same story of stalled bills and unfolding health disaster was repeated in the 1960s, leading in 1962 to new and important amendments to the 1983 Act. The adoption of these amendments resulted from several years of initiative for regulatory reform by Senator Carey Estes Kefauver. His intention was to institute compulsory licensing of drugs to increase public control over pharmaceuticals and the power of the FDA. His proposal failed to gain enough support and was soon withdrawn by the Senate Judiciary Committee. Three days after the bill was dropped from the Senate agenda, news broke out that the drug thalidomide (with marketing yet to be approved by the FDA) was the cause of the malformation of thousands of babies across the many countries where it had been sold (Lenz, 1992; Dally, 1998; Stephens and Brynner, 2001). The horrific malformations created shock waves and immediate support for increasing the FDA's power to check and possibly reject industry applications for marketing on the grounds of ineffectiveness (Daemmrich, 2004; Carpenter, 2010). Similar reactions occurred in other countries, leading to the generalisation of the regulatory instrument to prevent consumers having access to unsafe drugs: the licensing of drugs before they are distributed in markets or given marketing authorisation. This instrument also made its appearance in European regulation regimes, from the *visa* system in France (introduced in 1941 but revised in 1959 to address the consequences of thalidomide; see Chauveau, 2004) to Sweden's *Läkemedelslag* of 1961, UK's 1968 Medicines Act and West Germany's 1976 *Arzneimittelgesetz* (Champenois, 2001; Daemmrich, 2004; Hauray, 2006). The disaster also spurred the Commission of the European Community to initiate coordinating these regulatory reforms, proposing a directive in 1961 (eventually adopted as Directive 65/65)[2] that imposed on all countries of the European Community a system of pre-market licensing to check medicines against the three criteria of safety, efficacy and quality – a remarkable initiative just three years after the adoption of the Rome Treaty, that evinces the intensity of the thalidomide shock.

The legal creation of the European Medicines Evaluation Agency (EMEA)[3] in 1993 and of a centralised procedure for marketing authorisation and a system of pharmacovigilance is the distant consequence of the trail of regulatory reforms sparked by thalidomide. This regulatory reform was wrapped in a language of regulatory optimisation and member state coordination, for which various systems were tested and failed during the 1970s and 1980s (Hancher, 1990; Feick, 2004). It was justified by the feeling that marketing authorisation regimes were in crisis in the 1980s. The length of time it took to authorise a medicine was increasing at the very point when the complexity of products was growing and the competitiveness of European pharmaceutical industries was being challenged.[4]

The reform of pharmaceutical authorisation to create the EMEA is credited to a handful of experts and European Commission civil servants, reflecting probably the best possible way to combine the expertise of European countries into an efficient form of marketing authorisation, matching the level of complexity of upcoming innovative biotechnological medicines, the speed necessary for the European pharmaceutical industry to maintain its competitiveness, and rapid evaluation to clear the backlog of industry applications in national administrations. Two pieces of legislation were proposed at the end of the 1980s and adopted in 1993 after long negotiations between the European Commission, national governments and industry. They instituted a centralised procedure of pharmaceutical authorisation – in effect, a collegial evaluation by a nationally appointed member of a European committee, the Committee for Proprietary Medicinal Products (CPMP[5]), followed by a discussion by all national experts on this committee – and a European agency to formalise and propose decisions to the European Commission (rather than to make these decisions itself) (Kreher, 1997; Majone, 1997). The negotiations for these laws reflected a regulation regime in which national health ministries and agencies had great weight. The formalisation of the procedure of pharmaceutical marketing authorisation was riddled with issues on the division of labour between national experts and the CPMP, such as the list of products where the centralised procedure was mandatory, the selection of national delegates to the CPMP, and the staff and budget of the agency (Hauray, 2006).

The thalidomide disaster did not only lead to generalised clinical trials and the regulatory instrument of licensing. It also gave renewed importance to the instrument of information gathering in pharmaceutical regulation, via the invention of post-marketing surveillance or pharmacovigilance. The idea was to allow and entice physicians to report their suspicions of unsafe drugs.[6] When the state of a patient in treatment deteriorates, it may not appear evident initially that a medicine may be responsible for it. Without careful analysis of the patient's health, of the drug's profile (in information distributed with it, in drug dictionaries or in the medical literature) and the encounter between the two (when was the drug taken, how closely was the intake followed by the adverse event, etc.), without a good medical knowledge of the health event itself and its likely causes outside the taking of a drug, an adverse drug reaction, a rare one in particular, cannot be proven. Physicians may have a prescience and suspicion of it though. The lesson drawn from thalidomide, specifically by the World Health Organization (WHO) and pharmacology professors in a host of industrialised countries, is that there should be a system in place to collect this knowledge, both to enable rapid response to adverse drug reactions to avoid fatalities and serious health problems, and for progressing in the understanding of the role of drug iatrogeny in morbidity and mortality. The lesson of thalidomide may be explicated in the following way:

> The thing about Thalidomide was that, actually, the fact that it was an epidemic was recognised months if not a couple of years before the link with thalidomide was recognised. Somebody knew that there were lots of cases of phocomelia.[7] Paediatricians in Germany were saying: 'God we've seen lots of cases, what's the cause for that?' Is there something in the environment? Again it was easy to spot this, because phocomelia is incredibly rare in the absence of a teratogen being given at a very specific day of the pregnancy to cause this. So essentially the principle behind the monitoring, a doctor may see the case, might or might not see the link, but if you put the thoughts of lots of doctors together, and if they were able to report that, then . . . That's the principle of reporting that comes from thalidomide.[8]

It took a long time for all countries to understand this concept. It also took a few years to organise such systems of reporting and analysis of adverse drug reactions at the national level. Systems of vigilance and reporting were set up by the industry and the medical profession in various countries with support from the WHO (Lechat and Fontagne, 1973). The United States, Sweden and the United Kingdom were the first countries to set up such systems of reporting.[9] National centres of collection of adverse drug reactions were created (Lindquist, 2003).[10] In some places, the system made reporting legally obligatory for industry and health professionals. The standards that allow it to function – dictionaries of terms, formats for reporting, protocols for reporting and analysis, methods of causality assessment – were also progressively elaborated by medical associations from the 1980s onwards. Pharmacovigilance was relatively neglected in the original European pharmaceutical policies (Abraham and Lewis, 2000).[11] In general, it was not a factor in national projects to create autonomous regulatory agencies for pharmaceuticals authorisation. The Europeanisation of pharmaceutical regulation came from the creation of common procedures of marketing authorisation and not through projects of imputation of adverse drug reactions. Pharmacovigilance was a marginal topic in discussions between the European Commission, national governments and the industry.[12] Harmonisation between national systems of monitoring still progressed thanks to meetings within the pharmacovigilance working party set up by the European Commission (Commission, 1989). The adoption of EU laws for regulation of pharmaceuticals in 1993 nonetheless made 'pharmacovigilance' (a term preferred by the French school of drug safety, see Bégaud et al., 1994) a fully regulatory process, including obligations for the industry (to notify cases of serious and rare reactions within 15 days, to appoint a responsible and qualified person for pharmacovigilance) and leading to revisions of product marketing authorisations. The EMEA was also given the mission to collect and review signals. It progressively increased its work in this area, pushing for the centralisation of national data as well as supporting the work of the pharmacovigilance working party, which was integrated into its structure after 1993.

Since the combined impact of the worldwide withdrawal of wonder-drugs like Lipobay and Vioxx (respectively in 2001 and 2004, see Chapter 4), pharmacovigilance has taken a whole new importance in the agenda of drug regulators globally. The European Commission has now proposed a new regulation to codify all procedures of reporting and assessment towards further integration of national pharmacovigilance systems. It also makes clear that the primary obligation to monitor and report adverse drug reactions rests with the manufacturer.[13]

The regulation regime for drugs is thus structured by the marketing decision and two modes or tools of evaluation for all possible products before and after the marketing decision. Pre-marketing controls include 'scientific advice',[14] review of toxicological and clinical data, and delivering of marketing authorisation. Post-marketing surveillance involves the collection of signals of adverse drug reactions, request and review of company-sponsored post-marketing pharmacoepidemiological studies (the so-called 'phase IV' studies) and continuous changes to the terms of marketing authorisation depending on the accumulated information. Marketing authorisation and post-marketing surveillance or pharmacovigilance represent two phases in the life cycle of the product that is entirely shaped by this suite of controls. They are specific to pharmaceutical regulation and apply to all products – human and veterinary medicines as well as chemical and biological ones.

In contrast, food regulation has developed in a piece-meal fashion and with no overarching principle to guide regulatory developments (Bigwood, 1964; Smith, 1992; Macmaoláin, 2007). Before the era of food safety, food law tended to develop through 'vertical directives' (O'Rourke, 2005), corresponding to different types of products or segments of the food chain. The general food laws passed in a variety of industrialised countries in the early twentieth century (much reshaped in the early 1980s to reassert the liability of food producers for contaminations) mainly concerned inspection and enforcement (Paulus, 1974; Jukes, 1993; Canu and Cochoy, 2004). Until then, food legislation placed the fight against fraud and consumer deception above health priority, even if the latter was not absent from legal motives (Bigwood and Gérard, 1967; French and Phillips, 2000; Stanziani, 2005).

In 1999 though a rewriting of food law was inaugurated, with an overarching goal – consumer health and food safety. It took a political crisis, that of the BSE affair, for food law to integrate the notion that food production and distribution could be linked to uncertain adverse situations. The BSE affair was not a disaster per se. The number of fatalities does not make it a unique affair.[15] The BSE case is unique in that all ingredients for a political crisis were present, which would fuel, in the long run, tensions around the very credibility and legitimacy of European regulatory efforts in general (Majone, 1999), and around the way food issues are handled in particular (Clergeau, 2005; van Zwanenberg and Millstone, 2005; Ansell and Vogel, 2006). The BSE issue is one in which the challenge of uncertainty and its remediation by scientific research could be felt both materially and politically. The prion and the transmission of BSE to humans were a recent discovery, still plagued by uncertainties as to its vectors and impact (Granjou and Barbier, 2009). It also engaged essential differences between the science and expertise regarding veterinary diseases in France and the United Kingdom (Joly and Barbier, 2001). In a context of scepticism and opposition to European institutions as well as growing distrust of governments, the inaction of agricultural administrations on the issue came as a proof of policy incompetence and failure, if not of nepotism and corruption (Buonnano, 2006). It also worked as a symbol of the dominance of economic and veterinary definitions of the BSE issue over a public health frame (Miller, 1999), the misplaced domination of experts in the European policy machine and of a poor model of technocratic decision making (van Zwanenberg and Millstone, 2005).

The undue influence of scientific experts on regulatory decisions and the lack of transparency of scientific advice emerged as political issues through the revelation that BSE could be transmitted to people, and that agricultural ministries and the European Commission had done little about it because the public health issue was minimised by scientific veterinary advisers. In fact, the BSE affair had reverberations on issues of the legitimate relation between science and policy at national, European and global levels (Millstone and van Zwanenberg, 2002). In the EU specifically, several national governments and the European Commission came under very strong pressure after various inquiries by national and European parliaments revealed the wrongdoings and shortcomings of agricultural administrations in relation to consumer health. The European

Commission (particularly its Directorate General for Agriculture and the Standing Veterinary Committee that advised it) were accused of being biased towards industrial interests and prioritising achieving a single market above public health. A thorough European parliamentary inquiry revealed a number of faults and obliged European Commissioners for Agriculture to solve this crisis with policy and institutional reforms. The creation of a European food agency was initially raised in 1997 to demonstrate the Commission's willingness to solve the errors it made in the handling of the BSE issue.[16] It was decisively launched in 1999, but only after the collective resignation of the European Commission in March 1999 under accusations of corruption. In June 1999, the nominated Commission President Romano Prodi pledged to overhaul food law. As he disclosed his policy platform before the European Parliament he also announced the creation of a 'food and drug authority'. Prodi put forward the fact that the Commission President should not be held accountable for contaminations in the food chain, such as the one that was unfolding at that period in Belgium (the so-called 'dioxin scandal', in which dioxin-contaminated oil concentrate intended for poultry and cattle feed was distributed in Belgium, the Netherlands, and parts of France). The proposal to create a 'European Food Authority' (the name initially proposed by the Commission) marks an important shift towards a 'science-based' food policy and clarification of respective responsibilities of scientific advisers and regulatory decision-makers. The report of the European Parliament BSE inquiry committee confirmed the existence of a rather rare political consensus among the European Commission, European Parliament and national governments to restrict the missions of the future agency to risk assessment, leaving responsibilities for risk management untouched in the hands of the European Commission (Demortain, 2008 and 2009; Roederer-Rynning and Daugbjerg, 2010). In the aftermath of the crisis the main imperative was to separate science from politics or to draw a line between risk assessment and risk management. In this context, the 'independence' of the agency emerged as an imperative to restore the credibility of scientific experts, their neutrality, objectivity as well as excellence. The obsession with independence[17] signalled a willingness to depart from the previous system of scientific advice, whereby scientific committees were embedded in a European Commission directorate, to innovate institutionally by creating a self-standing and arguably more visible and transparent body (Maasen and Weingart, 2005).

Shortly after Prodi started assuming office, the newly created Directorate General for Health and Consumer Protection published a White Paper on Food Safety in 1999 that laid out all necessary legal revisions to existing food legislation as well as the possible scope, missions and resources of a

European food agency. A new regulation was proposed and discussed in the Council of Ministers and European Parliament at the start of the year 2001, and adopted in January 2002. The resulting Regulation 178/2002 (the so-called General Food Law) emancipated food legislation from its logic of mutual recognition and qualification of 'Europroducts' (Alemanno, 2007) to indeed turn it into an autonomous branch of European law (Holland and Pope, 2004), with its own constitutive principles. The General Food Law is built on the assumption, revealed by the BSE crisis and aptly expressed by a national delegate during legislative discussions in the Council, that 'it is not because a product complies with regulation that it is safe' (Council, 2001). The consequence of the BSE crisis for food law and regulation is contained in this assertion. Qualifying products on the basis of present criteria of quality and safety will never suffice to protect from uncertain, future-occurring and unexpected health issues, the detection and imputation of which the law must attend to.

No product was out of the reach of this legislative overhaul, whether animal foodstuffs, novel foods and food supplements, genetically modified foods, food additives or food contact materials, and so on. In reality, most types of safety issues were addressed from chemical or environmental contaminations to food-borne diseases or nutritional imbalances, whether their causes originated in primary production, food preparation, distribution or even in consumer behaviour (the farm-to-fork approach). Most tools were already in place: labelling and food inspections, the setting of limits and thresholds for contaminants in foods, systems of marketing authorisation or positive lists, certification of producers and distributors.[18] From the 1980s onwards, the logic of pre-marketing evaluation and imputability grew increasingly significant in food regulation with more products considered to be carrying health risks – enriched foods, and later novel and functional foods. But in contrast to pre-BSE food law, the new EU food legislation placed much more emphasis on a series of tools that concern the evaluation of adverse situations, contaminations and diseases. The General Food Law in 2002 (Regulation 178/2002) eventually signalled the legalisation of risk evaluation. The law effectively endorsed this view of food as likely to be unsafe, and of food regulation as reflecting a limited and provisional state of knowledge. In the general food law finally accepted, a generic approach to safety issues was adopted through the term 'risk', to mean 'a function of the probability of an adverse health effect and the severity of that effect, consequential to a hazard' (Regulation 178/2002). According to the law, regulatory measures in matters of food 'must generally be based on risk analysis' and on risk assessment as 'a scientifically based process consisting of four steps: hazard identification, hazard characterisation, exposure assessment and risk characterisation'.

In 2009, the EFSA produced 636 opinions for the European Commission, European Parliament or EU member states (EFSA, 2009). These represent a significant increase both year-on-year since the creation of the agency and in comparison with the number of opinions that the highly stretched scientific committees of the European Commission managed to deliver before 2002.

The law thus works to anchor regulatory action, in particular the setting and enforcement of standards into an evaluation of the risks that draws on experimental knowledge and tests as well as a more grounded knowledge of the adverse situations that occur in reality and people's exposure to them. Besides turning risk analysis into a legal principle, this renovated food law codified and generalised systems of traceability, of alert, vigilance and epidemiological surveillance. The General Food Law gave the newly created EFSA a mission of data collection. In the same vein, the Rapid Alert System for Food and Feed forcefully put in place and operated by the European Commission without clear legal basis in the 1990s was granted more importance and resources during this period. Whereas the inherited logic was one of product surveillance, it is increasingly used as a repository of data on food contaminations to progress towards epidemiological evaluation. One last example concerns the surveillance of food zoonoses.[19] Among the many new proposals put forward by the Commission following the BSE affair is the reform of Directive 92/117 to a full-fledged European-wide system of surveillance of these diseases. It has since developed around the EFSA and the transmission of data from national governments to the European agency to produce an overall report on the prevalence and incidence of these diseases.

Even if food regulation is not as neatly organised around two major controls (marketing authorisation and post-marketing surveillance) as pharmaceuticals regulation is, it still illustrates the rise of the aim of imputation, along with the standards and systems that organise it. The same story can be written from the point of view of the research in risk evaluation.

THE EVOLUTION OF EVALUATION

Safety and risk evaluation is a working topic for many scientists in terms of scientific research or of professional missions of social protection or economic utility. The intention of pharmacologists, toxicologists and microbiologists when it comes to foods and medicines has originally been the evaluation of their safety by reference to an experimentally defined state of health. This 'progress of experiment' (Marks, 1997) is not uniform

though, and evaluative sciences are not unitary. Like any scientific discipline, they develop in a variety of research fronts and approaches. And,
parallel with the enhancement of surveillance and monitoring of adverse
health events – rather than products – in regulation regimes, these disciplines developed in the second half of the twentieth century an interest
in the surveillance and analysis of factors of risks, be it for medicines,
technological foods and substances used in foods or conventional food
contamination issues. I examine below the logic of safety experimentation
and of the surveillance of risks for the three disciplines of pharmacology,
toxicology and microbiology.

Experimenting Safety

Clinical pharmacology is officially defined as the clinical study of the
action of drugs on the human body in experimental settings. It developed as a medical science in the mid-twentieth century. It grew out of the
asserted need to avail of a methodology that could systematically evaluate
drugs. This was in a context of increasing numbers of substances being
marketed as having curing effects, but where it appeared difficult to specify
the conditions in which their therapeutic value was greatest and their risks
smallest (Dangoumau, 2002).[20] Clinical pharmacology displaced toxicology to become the dominant science in testing drugs (Schmitt, 1982).
Pharmacology and toxicology share many things, including the study of
the biological dynamics by which a substance is taken up and eliminated
in the body, the mechanisms by which it acts on the body, and the notion
of dose–response relationships. However toxicology, as the science of
poisons, does not consider the positive effects or benefits of a substance,
whereas pharmacology is precisely founded on the notion that there is a
continuum, or balance, between the efficacy and safety of a drug (Marks,
1997). A drug should thus be evaluated by reference to a state of clinical
health and patient profile that it is meant to improve, by reference to what
in toxicology was reframed as 'non-clinical' or 'pre-clinical' testing in the
system of phased experimentation that is typical of drug development and
evaluation (Beck, 1989). Until substances started to cause fatalities by the
hundreds, a toxicologist's descriptive knowledge seemed sufficient. To
quote Marks (1997) in his study of the birth of clinical trials and modern
drug regulation, most products looked at until the 1930s 'posed little
problem'. The toxicologists' knowledge of the composition of different
plants and substances as well as their approximate experiments on animals
helped to explain what was acutely toxic, that is, posed a risk of death. It
seemed sufficient to know which product was likely to be toxic and at what
dosage (Gallo, 1996; Chast, 2002). According to an influential European

clinical pharmacologist, there has been a qualitative shift beyond toxicity testing in the evaluation of risks since the major disasters of the twentieth century:

> If you would discuss with the Polish pharmaceutical industry as we do at the moment to push them to develop systems of monitoring of adverse drug reactions, in all good faith they would say: 'but our products do not have any adverse drug reaction'. Well you see, the French thought the same before thalidomide, the British thought the same, the Americans too. – Would there be other cases, before thalidomide, that could have created a similar awareness of the risk of medicines? – Well, maybe, but we could not know at the time! We did not know. See, medicines were rather rudimentary. And they were mostly undergoing toxicological tests. We could realise quite quickly whether they killed or not. But that's about all we knew.[21]

Pharmacology went beyond toxicology in aiming, in line with the professional ethos of medical doctors, to look at both the safety and the efficacy of drugs as therapeutic means. It also scienticised the study of drugs by arguing that this state of health and the impact of the drug on it should be reconstructed experimentally in well-controlled conditions, rather than depend on non-standard and unreliable impressions from physicians. Drugs should be evaluated by reference to baseline conditions of a patient. This evaluation should also be quantitative. Small numbers are insufficient and testing a drug on one patient says little about how the positive and negative effects of drugs are distributed in the wider population covering very different patients and states of health.

From these principles, all sorts of experiment standards are derived. Pharmacology deserves the title of science as it provides methodologies and criteria for systematic comparison of the effects of drugs on a given health condition in experimental settings rather than through empirical and historical clinical impressions. It equips the 'logic of specificity' that undergirds drug regulation, by which one particular chemical intervention, to be approved, must be proven and measured to act specifically on one condition and set of symptoms (Lakoff, 2007). The art and science of clinical trials provide the set descriptions, criteria and scales by which diseases can be characterised and intervention assessed and measured. The US movement of therapeutic reform, the source of experimental clinical pharmacology (Marks, 1997), had always been interested in developing and putting forward the scientific component of medicine and therapeutic evaluation. The chance component of a drug being a cure, the need to assess the therapeutic merit of the drug as well as its safety across multiple cases rather than just one, and the subjective bias of the clinical investigator inclined reformers towards integrating statistical knowledge. The standardisation of methodologies and protocols of testing – culminating in the clinical guidelines produced by clinical institutions and regulatory bodies – is there to ensure the experiments are reliable, replicable for every new drug and that their results are comparable. Clinical trials imply a close selection of patients (with no health problems that could interfere in the analysis of the drug efficacy), whose health profile and 'baseline' conditions are determined ahead of the introduction of the tested drug. Statisticians also inspired the technique of randomisation. This is when patients are randomly put into the group using the drug or the placebo – and the double-blind – when neither the patient nor the physician knows whether the former is being treated with the drug or a placebo to avoid conscious and unconscious intervention in the treatment.[22]

Experiments were also employed as a way of ensuring the safety of technological foods. Although less concentrated and causing less acute toxicities, they nonetheless are likely to cause health problems that are difficult to anticipate or witness empirically. The industrialisation of chemical development and appearance of synthetic chemicals affected food as it did medicines through the invention of synthetic food additives that over time replaced substances (like salt) that were added to food for preservation (Morris, 2003). Other changes in food production explain why scientists considered applying emergent experimental methods to foods or certain components of foodstuffs. One is the multiplication of transformed foods. Basic foods have been diversified into dozens of types of products with slight differences in their composition, labelling and marketing. The increasing use of food additives is part of this trend. The third major transformation is the appearance of new and specific foods for which new categories have had to be invented, such as food supplements, diet and health foods, and so on. The last two trends were critical to the beginning of food testing. Newly developed compositions, substances or types of foods have generated the prescience that in food may lie the source of long-term health disorders. Cancer is the health issue that has motivated the invention of formal methods of risk evaluation for food additives and other chemicals found in food such as pesticides or veterinary medicine residues and chemicals in general, the development of which accelerated after World War II.

While toxicologists were seen as providing a limited expertise of drug therapeutic efficacy and safety (the value of animal models for a nuanced understanding of drug safety in humans being at best uncertain), it is indeed they who have taken up the task of evaluating food products or substances used in foods. Food toxicology grew from the combined influence of food analytical chemistry (the capacity to isolate substances from foodstuffs), biology's first mechanistic studies (to understand the mechanism of actions of particular substances in the body), and pharmacology with statistics and experimental sciences to design tests and trials. In its more practical dimension, toxicology's development was spurred by legislation to control substances added to food, initially at the national level and in the USA (Brickman et al., 1985; Young, 1989; Groenewegen, 2002) and later at the international level (Jas, 2010).

The main medical and scientific concern spurring international developments was carcinogenesis that was suspected to be a result of human exposure to chemical substances through food and the environment. This assumption and ensuing methodological developments came from other leaders of the discipline including René Truhaut. They had the prescience that toxicology as a science and art rooted in medicine should expand to cover issues of measurement and protection from cancer and other forms of chronic toxicity that depart markedly from the poisons that traditionally were of interest to toxicologists (Molle, 1984). The evaluation of food additives and carcinogenesis through food is a challenging area to be in, because what is being followed and evaluated is well known (it is the chemical substance that a company wishes to commercialise and introduce in manufactured food products), but the experience of the safety issues it may be linked to and that will provide the background to set its value in safety terms, builds up slowly. With food additives, toxicologists are dealing with chronic toxicity rather than acute toxicity. People are not dropping dead on the street because they have ingested foods containing food additives. They potentially develop cancer because of them. If this happens, it is within the course of a lifetime rather than within the short window pharmacologists typically work on. The specific problem of food toxicology is that chronic intoxications are 'much more insidious, and generally appear without any warning sign' (Truhaut, 1976) because of the repeated absorption of low doses. These doses may over time provoke health problems. But the symptoms are slow to appear and highly varied. Foods and food additives are bio-available substances. This means they degrade, and the fraction of the substance that is eventually absorbed by the body may have little to do with the dose initially ingested. Evaluation was not just prompted by natural aspects, but also by the fact that banning these substances would not protect the public as industries would eventually find ways of circumventing the ban and market substances in other forms. The concept that is used to define the conditions in which an additive can be released, the Allowable Daily Intake (ADI), is thus a widespread regulatory convention crafted by toxicologists.[23]

Microbiologists and veterinarians also developed an expertise in food testing and safety experimentation. Foods are not pure products since they are complexes of various substances and other components. Some foodstuffs are by themselves poisonous, at least when consumed in certain doses. Production and distribution of food is also less controlled and the distance between primary sites of production and the consumer offers possibilities of contaminations by micro-organisms such as bacteria, viruses and fungi. Much like food toxicology with regard to toxicology and medical sciences, food microbiology is simply a branch of microbiology or science of micro-organisms. Modern microbiology (considered through the work of Pasteur or Koch at least), was not simply a result of the discovery of micro-organisms. Those were known since the invention of the microscope. It was even known empirically, but not theoretically, that thermal processing of foods protected them from certain contaminations. The genuinely new and specific claim from microbiology is that there is a link between the life of these micro-organisms and diseases (Amsterdamska, 2009). It proceeded from the attempt to impute existing diseases to bacteria and micro-organisms. The now famous Koch's postulates are precisely the rules that Koch established with regard to this imputation.[24] As a result of these rules, which cancel out uncertainty and controversies around cause and effect, it is easy to set standards for a given animal disease or a microbiological contaminant (Bell, 1993).

Like microbiology in general, food microbiology grew in relation to the industrialisation of food production and as an auxiliary of it (Béjot, 2002). It covers different streams, which did not play an equal role in its development: matters of preparation and conception of foodstuffs (e.g. through techniques of fermentation), concerns over food processing, preservation and mitigation of food spoilage with such techniques as heat preservation and canning (Marth, 2001). This meant that for a long time food microbiology was based on analytical identification of pathogens. These were experiments to understand their behaviour along with mathematical modelling of their growth, from which food-processing criteria could be derived to extend to a more empirical approach through testing of foodstuffs at the end of the production chain or on the shelves. The public value of food microbiology for protection from food contaminations and food-borne diseases materialised later, after World War II. Food microbiology soared and became a closer relation to food hygiene when molecular biology techniques accelerated the identification and isolation of pathogenic organisms.[25] With the knowledge of these agents, more food poisoning outbreaks became explainable. From the 1950s onwards, food safety (as in protection from food-borne diseases rather than hygiene in food production) took a much more prominent place in the discipline. This evolution found its expression in the rise of a more preventive approach, with the invention of food safety assurance systems and the ambition to derive food process criteria from epidemiological knowledge of the prevalence of different pathogens and diseases in the population (Mossel, 1995; Notermans et al., 1998; Panisello et al., 2000; Notermans et al., 2002; Griffith, 2006).

Detecting and Finding the Causes of Adverse Effects, Contaminations and Diseases

Clinical trials now embody the field of 'pre-marketing safety', as opposed to 'post-marketing surveillance' or 'pharmacovigilance'. Pharmacovigilance comes, like clinical trials, from pharmacologists. Pharmacovigilance is the result of a pragmatic lesson about the heuristics of drug safety: the tangible picture of the safety of a drug can only be known by monitoring and being alerted to the various effects it causes on real patients, because serious adverse effects may materialise only very rarely and are missed in clinical trials. There is a need to be vigilant about adverse drug reactions. And unless there is an infrastructure for gathering and analysing data on the distribution and use of medicines, drug safety cannot be assured. It should be treated with urgency and in a state of alert. It is not about experimental patients but about monitoring the use of drugs to save lives endangered by these very drugs. It is also about using the experiential knowledge of physicians in the field as opposed to specialised clinical investigators trained in statistics. Pharmacovigilance revives the tradition of 'collective investigation' (Marks, 1997), in that it relies on physicians in the field, at the level of practice, and on the gathering and cross-checking of their knowledge. And effectively, pharmacovigilance or post-marketing surveillance is the branch of pharmacology that took on the organisation of this reporting by setting up systems of communication between physicians and the centralisation of reports. It does rely on the possibility of being able to derive causes (or factors) of medical events from large amounts of data, and the regularities that statistical calculations on these data reveal; however it maintains a close connection with the exercise of therapy. Statistical and epidemiological studies are used to confirm causes established at a micro-scale through the study of the singular situation of a patient.

In much the same way as clinical pharmacology and clinical trials displaced toxicologists, pharmacovigilance by medical doctors displaced pharmacists as the professionals with the original responsibility for and expertise in the preparation and use of medicines. Pharmacists have in fact been absorbed in this area now and populate the activities of pharmacovigilance, but the latter are dominated by medical scientists who define its methodologies, concepts and all that relates to the art of interpreting the medical causes of an adverse drug reaction. Pharmacovigilance thus extends the project of clinical pharmacology to scientise medicine. It shares the basic foundation elements of clinical pharmacology as it is about finding out what are the most rational, effective and essential treatments to cure people as opposed to a constant search for new substances. This is in terms of the precise dosage, timing of administration, and a host of other parameters that define whether a substance or pharmaceutical preparation will be effective or a risk causing a deterioration in patient health. It has also the same concerns for the comparability and cumulativity of efficacy and safety knowledge being rooted in statistical thinking.

However, pharmacologists who established reporting systems have drawn a different lesson from thalidomide: this is a lesson that replicates epidemiology's focus on the reconstitution of regularities on the basis of data collected from the field, rather than produced through experiments. For all their capacity to statistically inform about the conditions for the best use of a drug (including conditions of unsafe use), clinical trials are unable to gather all possible adverse reactions of a medicine. Despite their sophistication, clinical trials cannot pick up rare events. As the conventional critique goes, clinical trials are controlled experiments as well as a narrow selection of clinically diverse reality. First, trials take place within a limited time frame whereas the use of a medication may extend over very long periods. Second, clinical trials include a small number of patients in contrast to the large population that may be exposed to the medicine concerned, and the biological variability across this population. Clinical trials are not designed to accommodate this variation because it is intentionally reduced by selecting healthy volunteers within a limited age range to minimise the interference of other factors in the determination of drug efficacy. Third, clinical trials are ideal conditions for the intake of the drug under the supervision of researchers and physicians who check the timing of this intake, control the interactions between the drug and other drugs as well as the food, diet, and so on. Conditions in real life are not always portrayed in these test situations and many adverse reactions may surface from the way in which people take their medicines. Fourth and last, clinical trials are thought to leave too little room for safety evaluation. They are designed to check when and where a drug is efficacious. Adverse reactions that occur on the way are not all registered and investigated.

Altogether, post-marketing safety is built on an extreme sensitivity to uncertainty, in its technical as well as more structural dimension (Schwarz and Thompson, 1990). It demonstrates the diversity of relations of medical doctors to science (Berg, 1995), and the preservation within this domain of post-marketing safety of an important role for professional judgement and the capacity for interpretation of individual events occurring to patients with drugs, as opposed to the purely quantitative and highly standardised practice of clinical trials. Pharmacovigilance specialists know that the calculation of the frequency of an adverse reaction in absolute terms is impossible, given that the number of prescribed and consumed medicines is never known with certainty. They also know that the data they obtain are imperfect and incomplete. Quality and completeness are entirely dependent on the way in which the information is reported by physicians, and on the willingness of physicians to do this in the first place – and under-reporting is the first and main limit of systems of spontaneous notification, which has justified moving towards epidemiological trials based on following large but delimited cohorts of patients (Edwards, 1997). Adverse reactions may be very common health events with numerous possible causes. Disentangling these causes to focus on the responsibility of a medicinal substance in the event is such a difficult process that post-marketing safety specialists appear culturally reluctant to blame drugs or to exercise too strong a clinical autonomy (Marks, 2008). The case of practolol has caused quite a controversy, notably in the UK (Abraham and Davis, 2006). A British specialist of pharmacovigilance uses this case to illustrate the complexity of imputation:

> Interviewer: Was it just one or two cases which illustrated the fact that spontaneous notification was limited?
> Interviewee: Well, I think the biggest, there were several major problems but the biggest one was practolol. It is a beta-blocker. And was used in hypertension. The original beta-blocker made by the same company, ICI, was relatively selective. Essentially there are two main types of beta, 1 and 2. You must not block beta 2. There are not quite a lot of selective beta blockers around. It is normal to use. This was the first selective. For reasons no one really explained, it caused a unique ADR [adverse drug reaction]. A syndrome, called oculomucocutaneous syndrome. One of the things which was unusual about that, most adverse drug reactions, serious ones, are syndromes which have a lot of other causes, or had other potential causes. Relatively common adverse reactions, like hepatitis. Hepatitis can be caused by viruses, chemicals . . . all sorts of other things can damage your liver. If you put a liver under a microscope, it can be quite hard to distinguish the damage. That is quite often the case. So very often in clinical practice, an ADR is part of what we call the differentiated diagnosis. But what was unusual about practalol, really only two or three examples, here is a syndrome, of which there is no other known cause. It had not been described before the mid 1970s. And it has just not been seen before this drug was put on the market. That's very unusual. Because it was a multi-system disorder (eye and skin toxicity, guts and lung affected, etc.), because it was an unusual disorder, people ignored the cause. If you were a doctor and prescribed that to a patient, you would maybe see one case. But you're not going to see several cases. So this is the basic scientific principle of monitoring.[26]

Pharmacovigilance is thus caught between an ambition to perceive accidents as happening in real life with a therapeutically meaningful image of the frequency and circumstances of their occurrence. But there is hardly any tool to enable reaching such an aspiration. It is sceptical and cautious towards randomisation and other techniques for increasing the level of control and reliability of experiments. In this respect, pharmacovigilance has a formalised and mathematical approach that contradicts clinical experience and the therapeutic relationship with the patient. Harry Gold (1971) had distinguished clinical pharmacology as a drug-oriented and experimental science interested in how a drug acts, and therapeutics as a patient-oriented practical art of improving health conditions and the disease process.[27] In contrast with early proponents of medical statistics such as Fisher, pharmacoepidemiologists (though yet to be called this) were attempting to combine the expertise of statisticians and clinicians. To that end, they formulated the rules of statistical thinking in intuitive rather than formal terms. The physician and epidemiologist, pioneer of medical statistics and contributor to the methodology of clinical trials, Austin Bradford Hill notably articulated a set of markers for a robust causality analysis (Hill, 1965; Doll, 2003).[28] Their definition of statistical thinking in medical research was to become a cornerstone of typical epidemiological methodology. Hill has been most successful in getting British physicians to adopt the principles of controlled experimentation (Marks, 1997). And post-marketing surveillance is a direct illustration of this as Bill Inman, who was the founder of the British system of spontaneous notification of adverse reactions (the yellow card scheme), was a student of Hill. This light touch approach towards statistical imputation is well illustrated by the algorithm elaborated in France to guide clinical assessors in the operation of imputation.[29] It helps to disentangle the many factors that could be the possible cause of an adverse event and make the most of potentially confusing information provided by physicians (Miremont et al., 1994).

In all respects, post-marketing safety is a more distinct programme of evaluation than that based on clinical trials and other pre-marketing tests. The two forms of logic – qualification and imputation – are equally involved in clinical pharmacology as a research field. See Figure 3.1 for a mapping of the field of clinical pharmacology.

The data set consisted of papers published in three dominant journals of the field of clinical pharmacology between 1980 and 2004. Figure 3.1 displays the 15% most frequent associations (co-occurrence) of themes in this literature. It shows with great clarity how much the field of pharmacology is polarised. Cluster 1 concerns pre-marketing clinical and non-clinical testing of drugs, while cluster 2 concerns post-marketing drug safety and pharmacovigilance.

Cluster 1 comprises the following methodological descriptors in

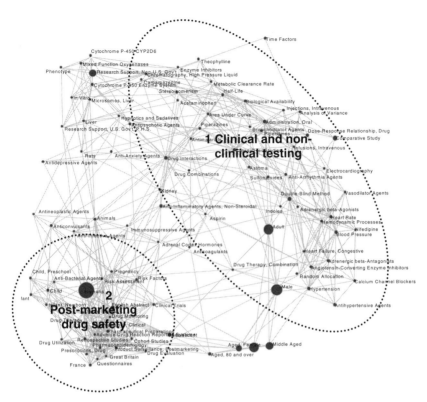

Figure 3.1   Co-word map of academic publications in the field of clinical pharmacology

experimental clinical pharmacology and clinical trials: double-blind method, random allocation, male, electrocardiogram, comparative study, dose–response relationship, and crossover studies. Cluster 2 concerns medical research on drug monitoring, ADR reporting system, retrospective studies, drug utilisation, cohort studies, pharmacoepidemiology, product surveillance, questionnaires, drug evaluation, risk assessment, and risk factors. This second cluster corresponds to pharmacovigilance. A comparison with cluster 1 shows pharmacovigilance is organised around a smaller set of descriptors as the links between these descriptors are shorter. This polarisation confirms that the field of pharmacovigilance has emerged slightly later than that of experimental clinical pharmacology. It is a literature that grew out of experimental clinical pharmacology and quickly opposed the latter for the lack of attention given to issues such as drug interactions,[30] NSAIDs,[31] aspirin,[32] female, middle aged (groups of patients generally not included in clinical trials (corresponding to descriptors that bridge the two clusters in Figure 3.1). The appearance of this area of research transforms the field of clinical pharmacology by polarising two areas that have been contrasted as pre-marketing and post-marketing safety.

Figure 3.1 also helps to visualise the distinct production in each of these two areas. The descriptors that form cluster 1 concern standards of experiments and tests: double-blind method, comparative study, dose–response relationship. These elements concern methodological approaches that may include a variety of different methods in practice to produce integrated measures of safety and efficacy in a way that makes different products comparable when evaluated on the same scale. Publications about specific tools and methodologies also appear more in this part of Figure 3.1, such as the 'analysis of variance' and 'random allocation'. Cluster 1 also includes descriptors of particular health conditions (blood pressure, heart failure, hypertension, asthma). Their presence in Figure 3.1 reflects the fact that much drug development is oriented towards these health issues. They are also important biological markers of drug safety used in clinical trials. Particular classes of drugs also surface here in cluster 1: anti-infective agents, anti-anxiety agents, anti-malarials.

Cluster 2 represents a type of drug research that is more generic, and oriented towards the development of heuristics to find out the most prevalent risks and impute them rather than a specific condition or class of drugs. Indeed, particular health conditions do not appear in this cluster. No particular drug or class of drug is referred to with the exception of anti-bacterial agents (and the presence of such an isolated term may be considered as an error). This is coherent with the history of the discipline as recounted by clinical pharmacologists (see Rawlins et al., 1992). The typical discourse of these specialists does not isolate iconic disasters. Instead, it looks at series of classes of products and types of adverse effects where management has turned out to be problematic for the profession and regulators, and fuelled scientific methodological developments.

> Interviewer: Do you see other important issues in the development of pharma-covigilance research?
> Interviewee: Yes, yes, absolutely. Thalidomide, Accutane, the non-steroidal anti-inflammatory drugs too [. . .] Take also the fluoroquinolones, that are still the object of lots of discussions. Cox-2 products currently still pose problems. And one should not overlook other products: anti-AIDS products, anti-cancer, treatments for haematological diseases, all of these are extremely difficult to analyse from the point of view of their adverse effects.
> Interviewer: So along your career there has been a handful of cases such as these ones that . . .

Interviewee: Yes, a handful of them. One should count them and do research about it, but there must have been around a dozen. A dozen of cases that really were outstanding and pushed to go further.[33]

An important implication of this is that there is a form of process of learning from incidents and controversies that appears continuous rather than episodic. The original thalidomide disaster inspired a state of vigilance, to be understood as a form of action but also as a kind of intellectual posture that recognises the open-endedness of a drug's lack of safety and uncertainty. Lessons are constantly drawn, methodologies evolve, and that is a situation that makes drug regulation less dependent on disasters than previously. Also of interest is the fact that the cluster is centred on one dominant node, humans, in comparison with cluster 1 that spreads around the descriptors, male and adults. This is a straightforward illustration of the fact that the approach of drug safety here is less mediated by clinical trials and the selection of patients that occurs therein. Pharmacovigilance looks at what it gets from physicians who report as well as from the cohorts of patients using a drug that it follows, and that are constituted to be representative of the whole population using the drug. Furthermore, the groups that are typically excluded from clinical trials are represented inside or close to the pharmacovigilance cluster (see for example the presence of such descriptors as child, infant, newborn, adolescent, women, aged over 80, pregnancy). Finally, it is more generic in its approach of methodologies and techniques of assessment. While cluster 1 includes terms that refer to rather specific methodologies (e.g. random allocation), cluster 2 looks at institutional systems of control (especially drug monitoring and product surveillance), which can take very different forms from one country to another, as the opposition between an epidemiology-based British school and a French school focused on individual assessment of reports from hospital and liberal physicians shows (Bégaud, 1993, 1999; Royer, 1993; Waller et al., 1996).

Food safety did not involve disasters as alarming as a pandemic of phocomelia, and certainly not one that could so unilaterally be attributed to one particular substance. But toxicology and food chemical safety still followed a path similar to pharmacology. The knowledge of food consumption and food safety in practice has rapidly been felt to be a frontier of toxicological research. Truhaut recalled towards the end of his life that 'because of the fluctuations in the type and the quantities of absorbed diets, the ADI is in reality an *integrated value with regard to the time* – a notion that is often forgotten' (Truhaut, 1991: 151; italic in original). By this he expressed the fact that evaluations such as ADI values do not reflect the reality of food contaminations, especially their distribution in the population and the relative risks that different groups face in relation to what they eat. As shown earlier, an ADI is based on experimental measures and animal models, much more than on an estimation of the reality of consumer exposure to foods and chemicals. The concern for the compilation of data about food chemical contaminations and food consumption grew during the 1980s and 1990s. Two toxicologists called for its development in 1984, at the occasion of the launch of a new journal on food additives and contaminants:

> It is necessary to develop analytical techniques to detect and quantify such compounds in foodstuffs often at levels so small as *to stretch* the most competent analysts; dietary surveillance data are needed to identify sources of potential hazard to the population as a whole or to particular sub-groups, and to evaluate the magnitude of exposure and the attendant risks. (Knowles and Walker, 1984: 1; italic in original)

National governments and food administrations have for a long time gathered data about chemical or microbiological food contaminations. For example, plans for the surveillance and control of contaminations of poultry flocks and foods with salmonella have existed in Sweden (a precursor country) since 1961 and the adoption of a law that aimed to avoid the repetition of a large-scale contamination that killed 90 people. Up to 30,000 samples are collected each year to check the prevalence and incidence of this type of contamination in the food chain. The United Kingdom having passed the zoonosis order in 1975, established a calendar and sampling plan to detect contaminations of salmonella (Dawson, 1992). France launched similar plans later in the 1980s, which were gradually coordinated in the 1990s by the European Commission and now the EFSA. Chemical contaminations have also been monitored by agriculture and consumer affairs administrations in most countries, through sampling and analysis plans. In Britain for instance, the Ministry of Agriculture ran a steering group on chemical aspects of food surveillance, which compiled food surveillance papers on the basis of recent analyses from the field (Lindsay, 1987; Knowles et al., 1991). Every country within the EU now has in place such systems of surveillance, which are inspected and coordinated by the European Commission through the Food and Veterinary Office.

Studies on chemical safety and microbiology, however, tend to look at data for contaminations occurring at production and distribution level, rather than food consumers and consumption habits (Peattie et al., 1983; Rees and Day, 2001). In the language of food chemical risk analysis, they have shown more interest in the monitoring of hazards than in the surveillance and imputation of risks, as shown in Figure 3.2.[34]

The map was produced using the table of contents of three journals of

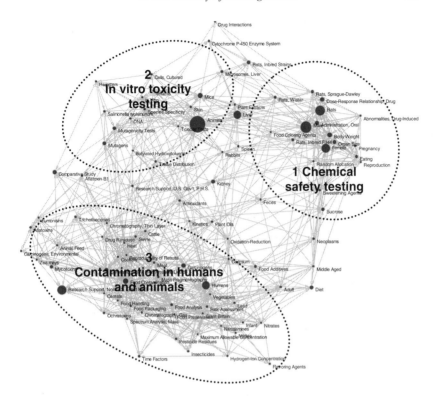

Figure 3.2   Co-word map of academic publications in the field of chemical food safety

the field, from 1980 to 2004. It also displays the 15% most frequent associations (co-occurrence) of themes in this literature. Cluster 1 (top right) concerns chemical safety testing, especially animal feeding trials, which is the main methodology for hazard identification and the current practice of most food toxicologists. Cluster 2 concerns in vitro toxicity testing. Cluster 3 is the area of study of food contaminations in humans and animals.

Cluster 1 comprises themes such as rats, drug dose–response relationship, oral administration, drug-induced abnormalities, body weight, organ size and random allocation. It is intertwined with the themes that comprise cluster 2, such as mutagens, mutagenicity tests, DNA, species specificity, in vitro, cultured cells. Cluster 3 is specifically concerned with the chemicals that are difficult to trace in production and in the environment and pose the most acute problems of contamination mitigation: aflotoxins, carcinogens, mycotoxins, drug residues, ochratoxins, pesticide residues, nitrates, and lead. It is about the technologies and techniques that should be improved to avoid contaminations and cover the following descriptors: food handling, food packaging, food analysis, food preservation.

From a dynamic point of view, the different themes presented in Figure 3.2 broadly fall into two periods (1984–1995, 1996–2004), between which the change in research interests seems rather clear. Food additives form one of the themes common to both periods, but cluster 3, for instance, comprises themes that all appeared during the latter period. This shows that the field was then concerned with the evaluation of risks to health posed by particular classes of substances, but failed to look beyond its preferred terrain to investigate new health problems. Where standards are concerned, this field has a large area devoted to technical norms for detection and analysis of the presence of chemicals in foods, as illustrated by descriptors like food analysis, mass spectrum analysis, gas chromatography, mass fragmentography. Thus as noted, toxicologists remain more interested in researching issues of detection of the presence of substances, than in approaches based on epidemiology that work backwards from adverse events to these substances as possible causes (Dayan, 2000).

The field of food and veterinary microbiology, like chemical contaminations of food, has not been reorganised around the epidemiology and causal analysis of food contamination and food-borne diseases. It remains structured by two disciplines: food hygiene and food protection (the latter is a study of food contaminations arising in food production) and animal health. This comes into sight more vividly in Figure 3.3.

Like the two previous maps, Figure 3.3 was produced using the table of contents of three journals in the field, from 1980 to 2004. It also displays the 15% most frequent associations (co-occurrence) of themes in this literature. The set of descriptors in cluster 1 of Figure 3.3 corresponds to the field of veterinary microbiology and animal health. Cluster 2 here relates to the management of food contaminations and food-borne diseases in the context of production. The epidemiological perspective is present in a range of descriptors in between the two dominant clusters. Those descriptors cover the research on systems of surveillance and collection of food contamination data.

Veterinary microbiology and animal health looks at the detection and curing of infections that usually affect primary food production and cattle rearing (bacterial antigens, mycoplasma, cattle, cattle diseases, sheep, sheep diseases, brucellosis, horses, horse diseases, animals, viral vaccines, vaccination, bacterial antibodies), with a strong relation to the molecular biology and in vitro analysis (cultured cells, DNA, RNA, molecular sequence data). Microbiology as an auxiliary of food production is interested in modelling of the behaviour of bacteria and other micro-organisms to predict the conditions in which they multiply and risk

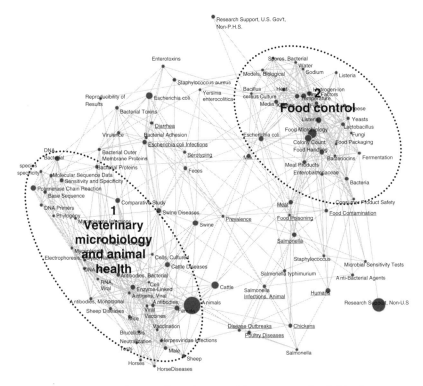

Figure 3.3   Co-word map of academic publications in the field of microbiological food safety

causing contaminations (see for example descriptors like colony count, biological models, heat, time factors, food temperature, kinetics). The descriptors such as yeasts, lactobacillus, fungi, fermentation also show continuity with the original microbiological questions investigated as part of the process of food conception and innovation. When it comes to safety, publications concentrate on most frequent hazards or causes of contamination: listeria, cheese, E. coli, meat products. The problem they pose is less one of imputation than of production. Causes of such contaminations are now known and there is plenty of experience about food contaminations and their control. The themes that bridge the two clusters include diarrhoea, E. coli infections, serotyping (the laboratory determination of the antigenic strains of the microorganism) on the one hand; prevalence, food poisoning, food contamination, humans, salmonella, animal salmonella infections, disease outbreaks, poultry diseases, chickens on the other. The research around the prevalence of salmonella is to understand more precisely which types of salmonella are common in animal populations, the cause of most food poisoning. These descriptors point to research to systematise sample collection in food contaminations, to analyse them to determine which particular serotype caused the contamination (there are nearly 2500 different serotypes for salmonella) and to maintain these databases to enhance the knowledge of disease prevalence.

The mapping of food toxicology and food microbiology research shows that the programme of imputation made only a marginal incursion into these fields and did not change their orientation (Gilbert and Scott, 2000). This is broadly coherent with the long-term orientations of these professions. Both toxicology and microbiology are sciences that, as far as food is concerned, grew thanks to industry's investments in technological developments, which these scientists supported in spite of the parallel roots of their disciplines in medicine (Claude, 1984; Malaspina, 1984; Amsterdamska, 2009; Demortain, 2010; Jas, 2010). Their dominant approach is one of qualification, of setting levels of exposure and fixed specifications following what they see as accurate and predictive experiments and tests. The standards of food epidemiological analysis and surveillance have come from parallel fields. The need to attend to risks – that is to adverse situations where control does not only depend on testing and experiments, but also on collection of data and information from consumers and businesses, and on thorny decisions about their priority as public issues – has however transformed food toxicology and microbiology in so far as these two fields have embraced the slogan of risk analysis and risk assessment. Toxicologists to start with have participated actively in the formulation of risk assessment frameworks notably in the 1970s in the USA. A long-lasting and influential framework was articulated in 1983 in a report by the US National Research Council, the so-called Red Book (NRC, 1983). Risk analysis emerged because there was uncertainty about the way scientific arguments were used to justify regulatory decisions that did not seem to logically derive from scientific data and analysis. After a series of contentious decisions by the US Environmental Protection Agency and the FDA to ban chemicals because of their carcinogenic risks (Jasanoff, 1990a and 1992; Boudia, 2010), Congress had indeed asked the NRC to assess separating the 'analytic functions of developing risk assessments' from regulatory decision making. The resultant Red Book codifies the four steps of risk assessment: hazard identification, hazard characterisation, exposure assessment and risk characterisation – incorporated exactly in this formulation in the European General Food Law and in the standards of the Codex Alimentarius.[35] Risk assessment is the combination of toxicology – interested in animal models and calculating dose–response relationships – and epidemiology. The committee put together by the NRC to produce this report indeed comprised scientists from both fields. Risk assessment has been seen as the product of an exercise of risk characterisation, which itself is a synthesised evaluation based on toxicologists' dose–response relationships ('hazard characterisation'), and epidemiologists' exposure assessment (gathering data and calculating levels of substances in food and diet, the overall amount of foods consumed, intake at the level of average individuals and in specific groups, etc.). The rise in risk assessment shows that food toxicology is being linked with the scientific enterprise of knowing more about the reality of food consumption and of food-borne diseases in order to reach differentiated rather than integrated evaluations of risks (Calderon, 2000; van den Brandt et al., 2002). Toxicologists have amended their approaches to include observational as well as experimental evaluation in ADI, the setting of which has been matched with the framework of risk assessment (Benford, 2000).

Some scientists from the field of food microbiology and food protection have moved to develop microbiological risk assessment. This is justified by the limits of food microbiology: incompleteness of data and imperfect understanding of biological processes in particular (Jouve, 1999). It is also a timely step to respond to demands from international trade bodies for formal and harmonised approaches to microbiological criteria that may serve as references in adjudication of a government's conflicts regarding the export and import of particular foodstuffs. In this perspective, predictive modelling appeared limited (Notermans et al., 2002), or insufficiently informed by disease trends and consumption patterns. The development of central epidemiological databases and coordinated food surveillance systems is a trend that has since strengthened. Human feeding trials were also used to generate data about human body infections following ingestion of different doses of a given micro-organism. This has brought about formal and quantitative dose–response assessments, the first of which were made by expert groups convened by the WHO (WHO, 2001). Chemical and microbiological risk assessment did not reorganise the whole field of food toxicology and microbiology respectively but only took some of the scientists and professionals from these fields in new directions. However, they still contributed to inform the law and practice of risk regulation as they generated standards for risk and turned the latter into an item of food regulatory law.

STANDARDS OF EVALUATION, STANDARDS OF REGULATION

This history shows that risk regulation is precisely about evaluating a product or activity with reference to the adverse events it may cause, based on tests and calculations, but also on experience of the adverse events, contaminations and diseases that prevail out there. Characterising these programmes of evaluation helps to emphasise a historical move from qualification to imputation, from safety evaluation to risk evaluation. Contemporary regulatory law and practice has turned much less positivistic. It has embedded product evaluation in the negative idea that there is no definitive certainty concerning quality and safety of products, given their intrinsic complexity, the ways they circulate, the conditions in which they are prepared and used, and their interaction and accumulation in the environment. Below, I aim to emphasise how the standards of each of these programmes gradually became law by going back to the description of regulatory regimes laid out in the first section of this chapter.

In his history of therapeutic reform in the USA, Marks emphasises that the 1938 bill (authorising the FDA to check the safety and composition of new drugs before they reach the market) acted on assumptions about drug safety made by therapeutic reformers. These assumptions, as noted above, included the notion that there was a trade-off between the efficacy and safety of a drug. A drug is unsafe when used for other purposes and in other circumstances than those in which it appears to be efficacious (Marks, 1997). The passing of this legislation represents a critical juncture in which the assumptions of scientists justified the establishment of ex ante control of drugs. From this point onwards, FDA officials have been able to assert a legal, more radical power over the regulated industry. This is also a result of the spread of scientists' conceptions of drug safety and the associated standards of experiment that were accepted as normal and appropriate even by the industry. The passage of a law and the emergence of a new type of regulatory check – pre-marketing evaluation – depend on the emergence of epistemic and practical conceptions by which the new legal provisions appear appropriate and fit as solutions to a problem, and which inspire regulators on how to communicate with the regulated industry: what sort of data to demand from it, how to interpret and work with these data to progress towards acceptable decisions.

Similarly, the progressive institutionalisation of pre-marketing evaluation for drugs since the mid-twentieth century can only be understood with the contribution of clinical pharmacology as it spreads particular assumptions about drug safety based on their therapeutic experience and experiments, and set technical standards for such evaluation. The chronology of scientific and legal developments shows that pre-marketing controls are not put in place, legally and institutionally, without the preliminary emergence and dissemination of particular epistemic assumptions about the safety of products and associated technical standards of evaluation. Regulation, in practice, depends on norms and conventions that define acceptable and appropriate data for checking safety and efficacy. Pre-marketing assessment in this sense took shape slowly, for instance through the norms and protocols of phased toxicological and pharmacological testing (Carpenter, 2010). In Europe, pre-marketing assessment will become more routine and will be extended to all drugs (not just new ones) once the science of toxicological, pharmacological and clinical testing has stabilised. Directive 75/318 represents the outcome of this inducement of science into regulation since it sets the standards and protocols for the different phases of product testing. The collective work of professionals in clinical pharmacology was instrumental in the institutionalisation of a practice of testing and evaluation – as therapeutic reformers were in the USA (Marks, 1997). It was enabled by the exchanges between medical experts of various countries, informally first and then through the scientific committees hosted by the European Commission since 1975 (Hauray and Urfalino, 2009), as well as the work of professional associations of pharmacologists.

Pharmacovigilance also emerged in the same way. Systems of reporting adverse reactions were initially private, as in the USA, where the collection of signals of adverse drug reactions began with an agreement between the American Medical Association and the pharmaceutical industry. This agreement was only later joined by the FDA to make it a public one. In the UK the chair of the Committee for the Safety of Drugs created in 1962, wrote to general physicians in the country in May 1964 to contribute to a system for reporting details on any negative patient conditions that might be imputed to a drug treatment. Bill Inman, a pharmacologist and flag-bearer of pharmacoepidemiology was asked to launch the reporting system known as the yellow card scheme in reference to the freepost yellow forms used by general physicians to return to the Committee information on suspected adverse reactions (Inman, 1993). A similar process took place in France with a centre first set up by the industry with support from a circle of pharmacologists around the country's university hospitals. It was later integrated into the health ministry, as a result of support from the Director for Public Health of the ministry (also a pharmacologist). These systems thus became public progressively through the 1970s.

Pharmacovigilance slowly took shape at the European level because of the gathering of these professors and researchers who were running national systems of pharmacovigilance in the 'pharmacovigilance working party'. In this CPMP sub-group created by the European Commission in 1975, leaders of pharmacovigilance programmes from member states met to review cases of adverse drug reactions that affected all of their markets, and to consider common rules of pharmacovigilance for the creation of a European-wide harmonised system of reporting. There was much to be done to approximate national approaches to the evaluation of adverse drug reactions in the market. The British school of post-marketing surveillance, inspired by epidemiology, judged its system according to its ability to assess the incidence and prevalence of adverse reactions. The French school has mostly been driven by the satisfaction of one central criterion, which is the ability to make precise causality assessments, even on single cases.[36] All of this had to be linked together to set up European rules ahead of the adoption of European post-marketing drug regulations.

What happens to pharmaceuticals applies to food additives and later, to the classes of products that toxicologists have selected to assess the risks of, such as novel foods. Food additives undergo a form of pre-market assessment that results in a decision to include a given additive in a 'positive list'. In legal terms, this list has the status of an annex to Directive 89/107, which sets the obligation for food manufacturers to use only those substances included in the list. This instrument does not equate with a formal marketing authorisation: anyone can produce and use an additive that is listed, whereas a marketing authorisation is granted to a particular manufacturer. But it still accords great weight to scientific evaluation of substances. This instrument of a positive list marks a shift towards considering such substances dangerous by default. Tests must prove how safe they are, to be included in the list. This contrasts with the earlier approach when the industry was allowed to develop and market substances that would be withdrawn if a lack of safety was established. In fact, the positive list was proposed by the international scientists who developed food toxicology and its main methodologies. This instrument was part of the so-called Godesberg Proposals named after the international conference in which scientists such as Truhaut and Druckley played such a major role (Jas, 2010).

The positive list system would be meaningless without a standard methodology to set the right norm for individual substances. The ADI has become this standard methodology. The procedures and criteria of toxicological evaluation of foods have been constructed directly in international circles, by experts such as the two named above, and particularly by a circle of toxicologists around Truhaut and members of the World Health Organization's Joint Expert Committee for Food Additives (JECFA). The JECFA was the site for scientific reflections around the standards of evaluation for food additives. Guidelines were discussed for protocols of experimentation and result interpretation. Most importantly, it was through the JECFA and international conferences organised by or around its members that the conventional methodology for setting safety criteria emerged. This concept was introduced by the JECFA in 1957 and is now used across jurisdictions, including in the FDA and the US Environmental Protection

Agency (EPA). There are a number of sister concepts such as the Reference Dose used by the EPA and many others. Put together, this conventional and integrated estimation of conditions of no-harm is the standard evaluative method for food components, additives, pesticide residues and veterinary medicine residues (Lu, 1988; Renwick et al., 2003; Galli et al., 2008), giving more weight to the eulogy of Truhaut by other toxicologists, that 'his most outstanding contributions have been in the application of his findings and those of others in the development of the principles that guide safety assessment and the application of scientific findings in the regulation of foods, food additives and other materials' (Chappel, 1984: 221).

As in food toxicology, microbiology food safety evaluation gained ground because of the international circles of experts who articulated the methodologies to systematise such forms of evaluation with microbiological criteria thereafter applied by national institutions. Comparable to the role of the JECFA in toxicity testing, the International Commission for the Microbiological Safety of Foods (ICMSF)[37] was established in 1962 to create the international criteria of judgement of microbiological safety, in parallel with national and European efforts. ICMSF objectives are 'to provide the scientific basis for microbiological criteria and to promote principles for their establishment and application; to overcome the difficulties caused by nations' varying microbiological standards and analytical methods'.[38] It proceeds by publishing books and papers that have circulated widely in the food industry, food hygiene and food protection profession as well as governmental and international institutions in charge of food safety policies. Food safety regulation, before the advent of risk regulation, was thus relying on the work of scientific committees or advisory institutions[39] to set and check microbiological criteria for different types of foods, starting with meat and meat preparations.[40] The setting of both the standard methods for microbiological assessment of food safety and uniform criteria has been heavily emphasised since the 1990s as necessary for the regulation of food safety and international trade. The importance of the ICMSF and of the science of microbiological risk assessment has been confirmed in the adoption of international specifications at the Codex Alimentarius (Debure, 2008; Winickoff and Bushey, 2010).

Epidemiological surveillance of food-borne diseases emerged like pharmacovigilance and pharmacoepidemiology as a form of self-organisation of veterinarians and physicians to put together their observations and cases, to detect epidemics. These networks were often first local, then became institutionalised nationally and, for some, internationally. Until the EU promulgated directives that harmonised data collection and surveillance, European systems could only progress because professors and researchers from various countries coordinated their work through common projects.[41] Directive 2003/99 formalised and juridicised these systems of surveillance that had initially emerged from scientists' and academics' self-organisation work.

Finally, risk assessment also leaves the hypothesis of a transfer from science to regulation as a plausible one. The four-step code for risk assessment diffused from the NRC in the USA to other international sites of regulatory science. Some scientists who participated in the NRC report were also members of the JECFA. Risk assessment specialists took part in the working group convened in 1988 to draft the Sanitary and Phytosanitary Agreement, as well as in FAO and WHO 'expert consultations'. Several members of the 1983 NRC committee were among those experts. During the 1990s, the four-step code of risk assessment emerged as an internationally accepted principle and became an instrument of the international trade regime (Horton, 2001). The reform of the GATT Agreement inaugurated in the mid-1980s consecrated risk assessment as an instrument for regulating trade. National decisions to block the import of a product have to be based on formal scientific methods that establish the probability of the hazard to the population. The Sanitary and Phytosanitary Agreement signed in 1994 made it an obligation for a country to effectively integrate the principle of risk assessment into national legislation.

The framework migrated into EU legislation via several channels. When the first BSE crisis broke out, civil servants and politicians took on the risk assessment–risk management discourse to make sense of what had occurred in order to design reforms. It quickly became natural to say, retrospectively, that the mishandling of the BSE disease (the first signs of the contamination of cattle appeared in the mid-1980s) was because of insufficient separation between the assessment and management of risks. In a dedicated risk-assessment unit within the Health and Consumer Protection Directorate, scientists acquainted with the common wisdom of the evaluation of chemical risks in food disseminated the risk-analysis concept in the European Commission. In parallel, other scientists sharing this vision involved their national governments in the framework. There has been a form of political acculturation to the language of risk analysis by regulators and scientists advising them, resulting in the original definition of risk assessment and risk analysis of the NRC report and of the Codex to be reflected in the General Food Law.

CONCLUSION

This chapter aims to show that regulation regimes and the concepts for evaluating risks evolved in parallel. It paints the landscape of risk regulation and prepares the ground for testing the hypothesis that, beyond parallelism, the adoption of these standards and systems was inspired by scientific developments. I thus extend the idea, present in Marks' study of therapeutic reform and other research, that the laws of risk regulation reflect the practical experience and assumptions of those researchers and professionals who experiment and observe risks. Not only are their assumptions carried forward in law. The standards that they develop also become part of the practice of regulation. The dynamic of risk regulation is not strictly one of response to disasters and crises. Obviously, they have an impact, but not on regulatory practice unless they are compared to other events or included in longer series of risks, with the result that they are translated into actual standards by scientists and professionals who proactively manage these situations and care for their resolution – sometimes well ahead of politicians, bureaucrats and public authorities in general. The historical overview also explains why regulation regimes for food and pharmaceuticals respectively are sometimes considered to be part of the same virtual domain of 'risk regulation'. Since they affect the evaluation of both food and pharmaceuticals, experience of structural uncertainty is seen to emerge that over time underlines the limitations of an evaluation strictly based on experiments.

Risk evaluation becomes inseparable from the production of regulatory standards as the programme of imputation grows stronger. Imputation is the true risk-evaluation programme, in the sense that it looks at the risks of occurrence of adverse events as results of decisions and interventions – of an industry to market a product, of product assessors to apply such and such test, of consumers to use the product at a given dose, time or in combination with others, and so on. In this sense, the programme of imputation is founded on a concept of risk as the risk of a decision, involving an intention to attribute dangers (Luhmann, 1993). This programme, therefore, is founded on an attitude of evaluating not only adverse events, but also human, industrial and regulatory decisions and behaviours. It is prescriptive, oriented towards the definition of actions (see for instance Raiffa, 2002) to avert these adverse events and replace faulty ones. Risk evaluation is almost by necessity an evaluation of regulation regimes and of their effectiveness. It is thus only normal, or this is the hypothesis at least, that from this programme standards emerge that apply to these decisions and behaviours. An imputative form of evaluation leads to the standardisation of control of risks.[42]

There are other complementary hypotheses emerging from this chapter. The first one is that as the programme of imputation increases in importance in regulating products, the alignment of more varied actors and sources of knowledge becomes a more critical challenge too. Contrary to the qualification of products, imputation relies on networks of actors that are more open and difficult to mobilise and order (Bodewitz et al., 1987), going from businesses to autonomous professionals and individual consumers and patients. All are made equally responsible to report information, and become part of the administrative and analytic systems (Buton, 2006; Marks, 2008; Langlitz, 2009). The era of risk regulation thus speaks to another regulatory order in which regulatory decisions cannot be prepared within close arenas and based on laboratory experiments, because everyone owns a share of the experience that allows anticipating and imputing uncertain adverse health events. Scientists develop standards that procedurally facilitate the channelling of all of these into collaborative processes of evaluation. But they may only be able to do so if they include all experiences found in their work of typifying and conceptualising adverse events and intervention. This implies that the expertise to set standards for imputation does not only originate in medical and biological sciences. It involves managerial and legal expertise too.

The chapter finally takes the first steps to show that there are differences between domains in this respect. Alert systems, vigilance and epidemiological surveillance take different forms for food or pharmaceuticals. They have also gained importance as regulatory instruments. Pharmacovigilance is a fully regulatory process since its actions are normally terminated by revisions of product specifications, and sometimes by product withdrawal. This is not yet the case in food regulation where detection, alerts and vigilance are only loosely coupled to the setting of standards. A preliminary observation here, to be confirmed by the study of particular regulatory concepts and colleges that shaped them, is that surveillance takes on a regulatory significance where it is organised by scientists who are also professionals – scientists with recognised expertise and legal responsibility in the distribution and use of these products. Subsequent chapters look in more detail at particular concepts, their trajectories and proponents to assess the extent to which science and regulation can reasonably be placed in a sort of equation, whereby the former will standardise the latter.

NOTES

1. Vogel (2003) specifically makes the argument that the EU became much more rigorous and precautionary in regulating health and environmental risks in the 1990s, succeeding in this to the USA, which had been following such a strong stance towards risks earlier in the 1970s. In this sense, US regulations can be said to have opened avenues for the development of regulation regimes elsewhere.
2. References to pieces of European legislation are abridged here as Directive or Regulation xxx/xxx for the sake of readability. Full references are at the end of the book.

3. Its activities only began in 1995. It is now known as the European Medicines Agency (EMA). I use the EMEA acronym for when it was in use before 2009. I apply the same policy to other acronyms that have changed, such as the UK Medicines Control Agency (MCA) becoming the Medicines and Healthcare products Regulatory Agency (MHRA) in 2004, the EU Committee for Proprietary Medicinal Products (CPMP) becoming the Committee for Human Medicinal Products (CHMP) in the same year, or the Heads of Agencies group (HoA) becoming the Heads of Medicines Agencies (HMA) in 2005.

4. National health ministries (in charge of authorising medicines in most countries before the creation of autonomous regulatory agencies) lacked the capacity to clear swiftly all incoming applications. Medicine agencies created in Europe from the early 1990s were meant to clarify a situation where health ministers were endorsing decisions from their own ministry without influencing much of the process. Two attempts were made to reform marketing authorisation procedures. Directive 75/319 allowed member states to permit without repeating tests and evaluation in their country a drug that had been authorised by another member state. In 1983, a multi-state procedure was also installed. These changes neither increased cooperation among member states nor convinced pharmaceutical companies of the benefits of using a coordinated or central European route for evaluation.

5. Now known as the Committee for Human Medicinal Products (CHMP).

6. The possibility for nurses to notify signals of adverse reactions started to be considered later in the 1980s in Sweden and in the United Kingdom. Discussions concerning the notification of adverse reactions by patients are also ongoing.

7. A congenital foreshortening of the limbs.

8. Interview with Patrick Waller, on 5 February 2004 in Southampton.

9. For more detailed histories of the post-marketing surveillance of drugs in the USA see Daemmrich (2004), Marks (2008) and Carpenter (2010).

10. Sweden being quite advanced in such systems of reporting, the WHO entrusted it with running an international drug monitoring centre, the so-called Uppsala Monitoring Centre.

11. And in general in the regulatory reforms that simultaneously took place across Europe in the 1990s (Hauray and Urfalino, 2009). The report that gave birth to the UK Medicines Control Agency for instance (Evans and Cunlife, 1987) did not consider the issue of post-marketing drug safety.

12. It was delegated by the leader of the reform, Fernand Sauer (head of the unit with this remit in the European Commission), to one of his team members. Sauer himself did not foresee pharmacovigilance at the time as more than quality assurance for marketing authorisation, an instrument to enhance the effectiveness and credibility of European decisions.

13. The manufacturer is required to maintain a pharmacovigilance master file for every product, a copy of which can be requested at any time by a regulatory agency. There are many other provisions in this proposal, which mark the full Europeanisation of pharmacovigilance and its recognition as a public instrument of control over products and their manufacturers (Arlett, 2004) in compensation for the limits to pharmacovigilance rules in the 1993 and 2004 reforms (Permanand et al., 2006).

14. Scientific advice covers the interactions between a pharmaceutical company and a regulator at the early stages of a drug's development to try to anticipate the tests and data that may be required during the official assessment of the product application. It has become institutionalised through the formation of a scientific advice working party at the EMEA and the establishment of procedures for it, as well as coordination between the EMEA and the FDA.

15. Despite initial estimations of several million victims of Creutzfeld Jacob disease following on from the transmission of BSE (Thomas and Newby, 1999).

16. That year, Jacques Santer, then President of the Commission, proposed with support from members of the European Parliament redesigning the Commission food safety

inspectorate into an independent agency (Valverde et al., 1997). Before proposing such an agency, the Commission sought other solutions along the lines of its committee-based food regulation. In February 1997, Santer proposed a two-fold reform. It was decided to transform the inspectorate into a separate unit of the Directorate General for consumer protection (DG XXIV at that time). All food-related scientific committees were transferred to the same DG. The assessment of risks and of consumer health by scientific experts became a key 'instrument' (Commission, 1997). However, these reforms proved insufficient. The committee system was unsustainable as experts were given too much work while the Commission had problems coordinating the activity of overloaded committees. Scientific experts began to ask for a more permanent body with more staff. The agency proposal thus resurfaced, this time as a solution to reorganising scientific advice. The commissioners for consumer protection and for agriculture raised this idea once more in 1998, during a hearing at the European Parliament.

17. Not a meaningful term since the European Food Safety Authority (EFSA) is organically linked to the European Commission, which retained risk management powers and shared much of its agenda with national counterparts.

18. The first directive on colours had been passed in 1960. Two years earlier, the systematic scientific evaluation of food additives had been imposed in the United States with the Delaney clause of 1958, according to which an additive that carries a risk of cancer shall not be marketed. The European Community began to regulate this area through Directive 1964/54 on preservatives, an official recommendation on tests for the safety evaluation of additives and finally, Directive 89/107 that created a positive list of authorised substances.

19. Diseases transmitted via animal source foods.

20. Harry Gold, a clinician from Cornell Medical College who allegedly coined the English term in the early 1950s, and Walter Modell a pharmacologist from Cornell University contributed greatly to the definition of this science, its aims and methods. Besides their input in the design of randomised clinical trials, they founded one of the main journals of the discipline, *Clinical Pharmacology & Therapeutics*. Pharmacology thus emerged in the USA, the UK and Sweden. The first university chair in clinical pharmacology was created in the latter country in 1940. In Europe, Paul Martini is seen as a precursor of clinical pharmacology (Shelley and Baur, 1999). It is now conventionally admitted that the evaluation of the therapeutic efficacy of drugs in humans covers clinical trials, pharmacodynamics (biological effects of the drugs on the body), pharmacokinetics (metabolism, dosage, fate of the drug in the body), pharmacovigilance (collection and assessment of information about adverse drug reactions after marketing of the drug) and pharmacoepidemiology (evaluation of drug effects and safety in large cohorts of patients).

21. Interview with René-Jean Royer, 16 January 2004 in Brussels.

22. Both techniques combined are constitutive of the randomised clinical trial as a gold standard of evidence-based medicine (Timmermans and Berg, 2003; Will, 2007). Clinical trials are also based on a notion of phased experiment inspired by FDA pharmacologists and oncologists of the US National Cancer Institute (Carpenter, 2010), by which knowledge gradually accumulated in toxicological testing, i.e. the non-clinical phase (then through clinical phases I and II) serves to specify the end points to further investigate in a randomised clinical trial (phase III).

23. The ADI is a conventional estimate of the dose of a chemical substance that a person may safely ingest on a daily basis through his or her lifetime without the risk of developing cancer. It is expressed in milligrams per kilogram of body weight. To set an ADI, the dose at which no adverse effect is observed (NOAEL) is first established on the basis of animal feeding trials. The ADI is obtained by dividing this dose by 100. The concept is thus a convention of toxicity testing and evaluation. It emerged within a circle of international toxicologists in which René Truhaut played a prominent role (Vettorazzi, 1987; Truhaut, 1991; Poulsen, 1995).

24. The four postulates are: the suspected organism is present in all cases of contamination and absent from healthy individuals; the micro-organism can be isolated in pure

cultures; the same disease occurs when the micro-organism is introduced in an animal; and the same micro-organism can be isolated again after infection of an animal.

25. *Listeria monocytogenes* and *Clostridium perfringens* were isolated in the 1940s for instance; since then *Bacillus cereus*, *Campylobacter*, Norovirus and prions have been identified as food pathogens.

26. Interview with Patrick Waller, 5 February 2004 in Southampton.

27. Something that may be sensed in the name of the journal in which most French pharmacovigilance experts have published and put forward as their journal, *Thérapie*.

28. These include strength, consistency, plausibility, specificity, temporality, biological gradient, coherence, experiment, analogy. They were for a long time considered to be the essential and exhaustive criteria. But there is now a tendency to revert to Hill's own caution in presenting these elements as mere guidelines or sets of arguments.

29. The algorithm, now a legal rule, stipulates that the analyst must first assess the case following a series of three chronological criteria (time between the taking of the drug and onset of the supposed reaction; evolution of the reaction after discontinuation of the treatment; evolution of the reaction and patient after re-administration of the drug) and three semiological criteria (rating the plausibility of a relation between the medicine and the observed clinical signs). These two sets of criteria lead to an synthetised mark of intrinsic imputability, which finally gets compared with the extrinsinc imputability, assessed through a study of the medical literature on similar health reactions.

30. Clinical trials are designed to study one molecule in particular, and provide little insight into the reactions that can arise from the intake of that drug and another.

31. Non-steroidal anti-inflammatory drugs: a class of product containing some of the most profitable drugs, where adverse effects have been verified through surveillance since the 1980s. The withdrawal of several of these products in the 1980s created controversy around the class of products as a whole (Snell, 1986).

32. Aspirin is a special topic in pharmacovigilance because the data accumulated since it came into use at the turn of the twentieth century have revealed benefits that were unforeseen then.

33. Interview with René-Jean Royer, 16 January 2004 in Brussels.

34. In the official language of risk analysis, a 'hazard' is the negative event associated with the consumption of a substance. The 'risk' is the probability of occurrence of that event. There have been extensive discussions around these definitions despite their apparent simplicity. 'Hazards' could as well be defined as well-established risks, causalities that are known, hence easily monitored because detectors are in place where they are known to occur. 'Risks' would then be about unclear causal relations.

35. The Codex Alimentarius Commission is an intergovernmental body under the joint authority of the World Health Organization and the Food and Agriculture Organization. It is tasked with the elaboration of international food technical standards. Compliance with the standards of the Codex have become mandatory since it was recognised as a reference body in the framework of WTO agreements in 1995.

36. These choices were directly reflected in the collaborative protocols used by agents to organise pharmacovigilance. France has established a network of regional centres with the specific ambition to stay as close as possible to doctors. Britain adopted a form of centralised pharmacovigilance instead, whereby signals are directly sent to a central office. The agency prides itself on having one of the most comprehensive databases of adverse events and of being able to run daily statistical calculations. In the French case, particular care has been put into the design of protocols that guide the evaluation of signals in the regional centres. A decision tree guides that evaluation and attributes a score to each drug depending on the answers to specific and systematic questions (e.g. Was the suspected product reintroduced? Has the adverse event occurred again after reintroduction?). The entire inputting process is closely monitored; junior doctors review the data sent in by regional centres under the supervision of a senior medical officer. Conversely, in the UK it is the procedures and criteria used to extract the key health impact data from the central database that have become especially sophisticated.

A protocol organises which data outcomes are to be considered by the doctors at the MHRA. In the case of significant health impacts, a whole procedure indicates who, when and how to organise the liaison with the drug's parent company, the depth of the risk/benefit assessment and the extent to which the relevant EU authority should be involved.

37. ICMSF was created by the International Union of Microbiological Societies to deal with food regulation and engage with the institutions that are responsible in this domain, starting with WHO. It is, in practice, a restricted working group comprising co-opted and internationally renowned food microbiologists. More will be said about this organisation in Chapter 5.

38. According to its website, www.icmsf.iit.edu/main/home.html, consulted on 15 June 2010.

39. These include the US National Advisory Committee for the Microbiological Safety of Foods in 1988, the French Centre National d'Etudes et de Recommandations sur la Nutrition et l'Alimentation, the Scientific Committee for Food in the European Commission or the Codex Alimentarius at international level, etc.

40. The relevant EU legislation was adopted gradually in the 1980s. Microbiological criteria are set by a number of directives, each of which covers a particular type of foodstuff, e.g. Directive 80/777/EEC for natural mineral waters, Directive 89/437/EEC for egg products, Directive 91/492/EEC for live bivalve molluscs, Directive 91/493/EEC for fishery products, Directive 92/46/EEC for raw milk, heat-treated milk and milk-based products, Commission Decision 93/51/EEC for cooked crustaceans and mollusc shellfish, Directive 94/65/EC for minced meat and meat preparations, Commission Decision 2001/471/EC for hygiene inspections, which itself enforces Directives 64/433/EEC and Directive 71/118/EEC on health issues in production and marketing of fresh meat – the first pieces of European food legislation.

41. The network of laboratories Enter-Net resulted from the initiative of two professors in public health from Britain and the Netherlands recommending a strategic development plan for surveillance to the European Commission. And the collective COST 920 programme is another example of scientists from across Europe developing a more systematic surveillance of a variety of food pathogens across the food chain, beyond the focus on particular micro-organisms in primary production.

42. This argument resonates with the demonstration by Desrosières that statistical thinking and epidemiologies develop along with the construction of regularities and a range of commensurable things, on which action is rendered easier. Imputation is, like statistical reason, a blend of cognition and action (Desrosières, 2002).

4. Drawing lessons: medical professionals and the introduction of pharmacovigilance planning

In 2004, the International Conference on Harmonisation (ICH)[1] adopted a guideline for 'pharmacovigilance planning' (PVP) that responds to the failure of the pharmacovigilance system to act on the discovery of unexpected, serious and sometimes frequent adverse effects associated with volume-selling drugs, and leading to a growing number of cases of worldwide withdrawals since the end of the 1990s.

Pharmacovigilance planning is one of the latest innovations introduced globally in response to the shortcomings of the monitoring of drugs and their adverse effects in the market. Controlling the risks posed by medicines is a never-ending task. In spite of methodological sophistication, clinical trials are always insufficient to uncover all potential adverse effects. The same applies to the monitoring of adverse drug reactions that were unknown at the time when the drug was authorised. Over several decades, various ways of collecting data directly from physicians, pharmacists and patients have been devised. Large sets of data are now accumulated about the number of products prescribed to a patient, when and how they were taken, and which effects appeared. Statistical analyses are made on these data sets to try to find warning signals of a drug directly causing a negative health event. This information is always limited in some way, notably because businesses and physicians fail to report a lot of the cases they see. Given these limits, pharmacovigilance is prone to failure – and to constant improvements.

The idea of pharmacovigilance planning crystallised soon after 2000 and a series of abrupt withdrawals of drugs from the market.[2] These radical decisions proved how difficult the management of subtle shifts in the risk/benefit balance of drugs was, and that competences and tools were missing to make sure medicines could be targeted to the groups of patients who would benefit from them, and only to those groups. PVP is about anticipating the situations where risk minimisation and risk communication measures must be implemented, often urgently. The concept states that provided the state of knowledge about the adverse reactions linked to a drug, proven and unproven, is specified when this drug enters the market (the pharmacovigilance specification) and that measures of risk management are established in advance (the pharmacovigilance or risk-management plan), then the likelihood that risks can be better managed and that withdrawal decisions will be avoided is higher. PVP is neither about pre-marketing safety nor about post-marketing surveillance. It is about anticipating surveillance at the stage of marketing, about being prepared to implement measures of communication to doctors and patients, of labelling changes. PVP directly remedies the lessons learnt from these abrupt withdrawals, such as the lack of continuity in the actions taken to minimise the occurrence of adverse drug reactions; lack of procedures to adjust regulatory decisions in complex risk/benefit assessments; ensuing failures of regulators to make decisions before manufacturers to prevent the full withdrawal of products by manufacturers. PVP, therefore, has been designed as a set of operations and information formats for manufacturers to establish what is known and unknown about the safety of their product as it goes on the market and pre-design risk minimisation actions. It provides ways for pharmaceutical companies to better plan their monitoring activities, and decide on which pharmacoepidemiological studies to make in the future. It is also meant to ease communication between companies and regulatory agencies.

The PVP case is interesting because it illustrates the rapidity with which new international harmonised standards are effectively put in place in medicines regulation to reform its failures. Only four years elapsed between the first appearance of notions of specification, planning or preparation among academics and practitioners of pharmacovigilance, and the subsequent adoption of ICH and EMEA guidelines based on them. What is more, these guidelines sparked new processes for drug safety in the industry and regulatory agencies. Pharmacovigilance and risk management plans are now routinely submitted by the industry to regulatory agencies. Licensing and post-licensing units in regulatory agencies have established common routines and cooperation. These changes are all the more remarkable as they contradict the regulation of medicine safety as it developed historically. The 'planning' activity is a bridge between two organisational and professional worlds – pre- and post-marketing drug safety – with no history of cooperation. However shocking, drug withdrawal affairs do not by themselves explain this fast and historical reunion of two opposing professional conceptions of drug safety control. The PVP success is explained by the circulation of expert drug safety scientists, from whom the more specific PVP concept originates, among other regulatory actors and their alignment on a risk-management agenda. This chapter follows these scientists, all of whom are doctors, who used the

ideas present in a broader transnational circuit of expert drug safety scientists to produce PVP, but also between the latter and major regulatory
agencies engaging with the risk-management agenda. In sum, the PVP
case reveals the symptomatic evolution of drug safety experts towards a
greater embeddedness in regulation, and the accelerating effect on scientists' standardisation work created by industry and regulators' pressure to
harmonise.

PLANNING SURVEILLANCE

The work on harmonising guidelines for pharmacovigilance planning
started in the ICH in September 2002. Two years sufficed to conclude the
work and issue the guideline ICH E2E, with immediate impact on the
practice and organisation of pharmacovigilance activities in the industry
and in regulatory agencies (ICH, 2004).

The resulting guideline establishes that manufacturers should provide,
at the time of licensing, the knowledge they have about the toxicity and
safety issues that clinical trials have shown, the safety issues that may be
suspected based on knowledge of similar drugs but were not flagged up in
clinical trials, the populations that were included in the clinical trials and
those that were not, what is known of negative drug interactions, and the
hypotheses that can be made about other detrimental drug interactions on
which no information is yet available. The guideline is an incentive to explicate all knowns and, most innovatively, 'known unknowns' on the safety
of a product. These will constitute what is called a 'pharmacovigilance
specification': a risk assessment criterion for those in pharmacovigilance
to determine areas in which adverse reactions are likely to emerge, and a
way to quickly sort out expected from unexpected events. The guideline
also seeks to establish, beside the pharmacovigilance specification, a 'pharmacovigilance plan'. This should contain the actions that can be taken on
each safety issue, the methods of surveillance or pharmacovigilance that
may be used, the epidemiological studies that may be launched, and so on.

The PVP guideline made its way into the EU regime for pharmaceuticals almost instantly. The European code for pharmaceutical products
(updated by Directive 2004/27) and Regulation 726/2004 stipulate that
applications must comprise 'a detailed description of the pharmacovigilance and, where appropriate, of the risk-management system which the
applicant will introduce' (article 8). The idea of a risk-management plan
or system is the translation chosen by the EU for the ICH guideline on
PVP. In 2005 indeed, the European Medicines Evaluation Agency created
a guideline for the submission of risk-management plans based on the ICH

guideline (EMEA, 2005), incorporated in the book of European official guidelines on Pharmacovigilance for Medicinal Products for Human Use (the so-called 'Volume 9'). As in all such guidelines, the risk-management guideline is extremely detailed and prescriptive:

> The Safety Specification should be a summary of the important identified risks of a medicinal product, important potential risks, and important missing information. It should also address the populations potentially at risk (where the product is likely to be used), and outstanding safety questions which warrant further investigation to refine understanding of the risk–benefit profile during the post-authorisation period. (EMEA, 2005: 8)

An extremely detailed list of information items follows, illustrating the extent of the resources and data systems to be mobilised by the manufacturer. The safety specification should be based on information concerning the non-clinical part of the safety specification (toxicity, general pharmacology, drug interactions, other toxicity-related information or data), on limitations of the human safety database (e.g. related to the size of the study population, study inclusion/exclusion criteria; with particular reference to populations likely to be exposed during the intended or expected use of the product in medical practice), the populations not studied in the pre-authorisation phase (children, the elderly, pregnant or lactating women, patients with relevant co-morbidity such as hepatic or renal disorders, patients with disease severity different from that studied in clinical trials, sub-populations carrying known and relevant genetic polymorphism, patients of different racial and/or ethnic origins), adverse events/adverse reactions (the important identified and potential risks that require further characterisation or evaluation), identified and potential interactions including food–drug and drug–drug interactions, the epidemiology of the medicine (incidence, prevalence, mortality and relevant co-morbidity, and should take into account whenever possible stratification by age, sex, and racial and/or ethnic origin), the pharmacological class effects (risks believed to be common to the pharmacological class), as well as a few 'additional EU requirements' (potential for overdose, potential for transmission of infectious agents, potential for misuse for illegal purposes, potential for off-label use, potential for off-label paediatric use). All of these data requirements imply close interaction between the different parts of a company that hold these pieces of information. The guideline also de facto requires the company to enter into close interactions with regulatory agencies, to discuss when and where to submit a risk-management plan, when and where to include 'additional pharmacovigilance activities' or indeed to decide on the timing of these discussions.

In this guideline, a third section was added to the pharmacovigilance specification and pharmacovigilance plan of the ICH guideline: the risk minimisation plan. This specifies the measures that the company would put in effect if one of the safety issues materialised – such as production of educational materials, communication with pharmacists to ensure transmission of information to patients, change to the legal status of a medicine, setting up of a patient registry or of a restricted access programme. A risk management plan should be submitted with the application for a new marketing authorisation (for any product containing a new active substance, a similar biological medicinal product, a generic/hybrid medicinal product where a safety concern requiring additional risk minimisation activities has been identified with the reference medicinal product, advanced medicinal therapies) and with an application involving a significant change in marketing authorisation (e.g. new dosage form, new route of administration, new manufacturing process of a biotechnologically derived product, significant change in indication/patient population), unless it has been agreed with the competent authority that submission is not required.

As a result of this guideline, risk-management plans devised by manufacturers are now routinely examined and approved by EU regulatory authorities, at least for specific substances. The EU Regulation on pharmacovigilance proposed by the European Commission in 2008 aims to provide the full legal basis for requiring manufacturers to provide such risk-management plans for all newly authorised products (Commission, 2008b). In practice, and already before this last piece of legislation was adopted, the standard produced a number of changes, in particular under the heading of 'risk management'. In the industry, the adoption of an international guideline, the mutual expectation between industries and regulators that more planning will be requested and done, the increase in the number of professional and medical publications on risk management, the proliferation of training on risk management directed at industry and regulatory safety officers and thus of competences, and the improvements in the quality of company-sponsored phase IV studies (Stephens, 1994), all mean that the concept quickly produced pervasive effects on organisational structures and processes (Andrews, 1997). A complex system is set in motion comprising training bodies, drug information companies, conference organisers, regulatory affairs journals, which will relay the new concept towards a self-fulfilling end.

PVP seeks to solve one of the old limitations of drug safety systems. As exposed in Chapter 3, the safety of drugs before their placing on the market is mainly verified by clinical trials. The controlled conditions in which these trials take place typically offer the illusion that safety can be evaluated in a rather absolute manner. Thalidomide and ensuing methodological research revealed that this was not so. Only the most frequent reactions are spotted. Less frequent – but also serious – reactions are missed. Pre- and post-marketing safety are contrasted by respective practitioners as an opposition between experimental and 'real-world' drug safety. For pharmacovigilance scientists, pre-marketing evaluators assess safety in quasi-virtual terms through the artificial conditions of clinical trials and the reporting of their results in data dossiers.

This opposition illustrates a particular order in regulatory practice, with a strong partition between pre- and post-marketing safety, both in the industry and in regulatory agencies. Pharmaceutical businesses tended in the past to relegate safety officers to dedicated units, which have always had difficulties in collaborating with R&D, marketing and clinical development units, and often failed to convince them to record adverse reactions occurring in clinical trials to anticipate the post-marketing phase. The legislation on pharmacovigilance, paradoxically perhaps, reinforced this effect by instituting the industry role of the 'pharmacovigilance responsible person' (legally responsible for the recording of signals sent by physicians, for the validation and transmission of these reports, the performance of surveys on the most serious cases, and causal analysis, as well as auditing of internal systems of monitoring) instead of joining up pre-marketing clinical testing and post-marketing surveillance. Adverse events emerging from clinical trials and from systems of post-marketing notification by health professionals were seldom centralised, let alone reported.

Similarly, the typical organisational chart of regulatory agencies features the two divisions, one for licensing and another for post-licensing. This separation has long been seen as functional and necessary for the impartial evaluation of product safety. Not having taken part in the examination of the data submitted by the manufacturer and in the decision to market the drug guarantees that the approach to data collected from users and prescribers is unbiased. Safety signals will be evaluated without risk of wanting to 'protect' the drug.[3] However, cooperation across this separation has frequently been difficult. A side-effect is that post-marketing safety officers know too little about the drug they monitor, where to search or what signs to be attentive to, being unable to build on the clinical and toxicological data thoroughly examined before licensing. Pharmacovigilance is typically reactive as it only looks at medicines when potential safety issues are signalled. If no signal emerges, the pharmacovigilance units of manufacturers and regulatory agencies have no reason to study the product. This means that months or years can elapse between the licensing of a drug and the assessment of safety signals. Knowledge would be lost if communication with those who studied the drug in clinical trials was cut. Pharmacovigilance people would then need to start their assessment from scratch when a signal arises, thus slowing down the process of reaction. PVP created an opportunity and provided guidance to change this.

Changes have concerned primarily the connection between drug development and drug safety teams within companies. The PVP guideline induces the participation of pre- and post-marketing safety people in a collaborative process of collecting, evaluating and acting on safety information, as well as a collection of adverse effects arising from clinical trials within the databases maintained by pharmacovigilance services (Hartford, 2006; Maennl, 2008). Changes are analogous on the side of regulatory agencies. Several European agencies including the EMEA organised the cooperation between pre- and post-marketing divisions, notably by facilitating the participation of a pharmacovigilance officer in pre-marketing evaluations. People trained in pharmacovigilance or with experience in post-marketing safety assessment bring different sorts of skills in these pre-licensing meetings, notably their capacity to deduce, from the posology, the structure of the molecule, the mechanism of action, the form of the medicine, its intended use, and which side-effects are most likely to appear from widespread use of the product. While they are generally seen as insufficient to block the licensing of a medicine, these safety concerns help to anticipate future issues and to realise that the work to assess and assure safety is not done and over once the product is marketed. Regulatory agencies also recommended letting companies meet pharmacovigilance scientists at the early stage of 'scientific advice' that precedes the submission of an application. They renamed their post-marketing departments 'risk-management departments', and centralised within them all surveillance services and databases as well as the new activities relating to the assessment of the protocols and results of pharmacovigilance activities foreseen in manufacturers' plans (AFSSAPS, 2009).

The main language used, in the law as well as among regulators and among drug safety specialists to account for all of these changes was that of 'risk management'. Pharmaceutical risk management breaks with the pre- vs. post-marketing opposition to emphasise the necessary continuity in the accumulation of knowledge and signals about the safety of the drug, from clinical trials to its early years in the market and later, to reach better reactivity and appropriate regulatory responses to adverse drug reactions. The risk-management language prevailed in the EU, where the PVP guideline was translated into an EMEA guideline for 'risk-management systems' and the EU legislation speaks of the submission by companies of 'risk-management plans'. In the guideline produced by the EMEA to integrate in EU standards the ICH one, 'A risk management system is defined as a set of pharmacovigilance activities and interventions designed to identify, characterise, prevent or minimise risks relating to medicinal products, and the assessment of the effectiveness of those interventions' (EMEA, 2005: 20). The pharmacovigilance and risk-management languages have thus been brought into line with one another. The PVP concept thus integrated a broader and indeed ambiguous agenda of pharmaceutical risk management, in such a way that people involved in product development and clinical tests, in pharmacovigilance, and in managing regulatory agencies, seemed to be aligned on the same ideas. How was such alignment produced and what had the concept of planning to do with it? To understand this, it is necessary to go back to the projects of reforming pharmacovigilance that followed the nearly catastrophic withdrawal of the drug cerivastatin.

DRAWING LESSONS FROM REGULATORY FAILURES

From Cerivastatin to ICH

In August 2001, the manufacturer of cerivastatin,[4] Bayer, decided to withdraw it from all markets worldwide. The product had caused the death of 100 American patients. Cerivastatin is a lipid lowering and lifestyle drug that belongs to the statins family. It is a 'me-too' drug, developed by Bayer to take its share of the turnover generated by this class of substance in a context of rising concern over cardiovascular risks. Cerivastatin was also a 'blockbuster' drug; it generated over a billion dollars of turnover per year. Like other statins, cerivastatin had been linked to muscular problems in clinical trials. Bayer's decision to withdraw the product was motivated by the surge of deaths from rhabdomyolysis (the rapid breakdown of muscle tissue, causing the potentially fatal release of muscle cells in the plasma and kidneys), caused in cases where this drug was used in combination with a lipid-lowering drug of the fibrate class (gemfibrozil). The withdrawal of the drug was not requested by regulatory agencies that, in spite of the fatalities associated with it, wanted to keep it on the market for certain groups of patients for which it was critically important as a therapeutic drug.

Cerivastatin was a regulatory failure, since the system of pharmacovigilance is premised on the possibility of being able to rationalise the use of therapeutic means in order to reveal adverse reactions that are not so much signs of an absolute hazard posed by the product as an invitation to specify further the conditions in which the risks exceed the benefits for different types of patients. It illustrated the incapacity of regulators to determine and target the groups of patients that more specifically need a drug, to avoid full withdrawal. Instead, a decision made on the grounds of profitability takes a potentially interesting means of therapy away from groups of patients who might need it in spite of the risks and that, what is more, are often badly informed about withdrawal decisions. Cerivastatin, furthermore, was a 'me-too' drug marketed by Bayer to take its share of the profit generated by a new class of cholesterol-lowering substances, the statins, and it has often been suspected that the additional therapeutic benefits of a me-too drug compared to others already on the market are insufficiently demonstrated. Cerivastatin led European regulators and health policy-makers to admit that it was time to improve the conduct of pharmacovigilance for a more comprehensive management of medicine risks.

The prevailing pharmacovigilance system was blamed in the press and by politicians for being insufficiently reactive to already known safety issues. The UK regulatory agency was especially targeted. Although it formally was 'rapporteur'[5] for this product, it did not react any more swiftly than other agencies. The UK and European regulatory agencies did have reports at their disposal signalling adverse effects involving ceriv-astatin. They did advise doctors and patients to avoid the fatal combina-tion of cerivastatin and gemfibrozil. The official 'Summary of Product Characteristics' had also been modified in July 2001. However, it appeared that Bayer had been notified of further cases of deaths in the few weeks pre-ceding its withdrawal and took the decision to remove the product without notifying regulatory agencies. A month after the withdrawal, Bayer pro-duced a report to explain its decision, in which 59 cases of death between November 1998 and June 2001 are listed, of which 40 deaths (mostly in the USA) had been signalled to the French medicines agency. Eleven cases occurred in the EU and notification was given to the UK regulatory agency in charge of the product within the framework of the European regula-tory system (AFSSAPS, 2001). Bayer failed to notify European regulatory agencies of eight cases of death that occurred outside the EU.

The cerivastatin affair impacted more than pharmacovigilance. The affair was symbolic of the negative effects of the whole architecture of the regulatory system for pharmaceuticals based on the partition between pre-marketing and post-marketing regulation. In particular, it illustrated the pressure for shorter times-to-market of drugs, and the suspicion that this pressure meant a reduced capacity to spot rare adverse effects before the marketing of the drug and more withdrawals afterwards (Abraham and Davis, 2005). The case also crucially illustrated the lack of diligence of companies sending sensitive safety information to regulators, despite their legal obligations, for instance, to send all signals of serious and rare adverse drug reactions to regulators within 15 days (Arlett and Harrison, 2001). Similarly, the lines of communication between regulatory agencies and health professionals seemed faulty. Other critics observed that there are no comparative tests to evaluate the added-value of 'me-too' drugs before granting marketing authorisations. These are drugs with similar pharmacological and therapeutic properties to already approved drugs. They are generally more expensive, bring little additional benefit to patients and may create new adverse events. Cerivastatin was allowed on the market just like other statins, whereas there was no robust knowledge of its therapeutic added value. The affair highlighted the tendency of the drug industry to be less innovative and to concentrate on the marketing of such products, particularly, 'life-style medicines'. Finally, it exposed the difficulty in articulating risk/benefit assessments and adjusting controls on the distribution of the medicine, as well as the communication to patients of withdrawals (Wolfe, 2002). The event thus came to embody the failure of pharmaceutical regulation and of pharmacovigilance for the first time in the history of this relatively neglected regulatory intrument.

The cerivastatin affair turned into a symbolic case for reform, in response to which the executive director of the UK MCA took a decisive initiative. He called one of his in-house post-marketing safety specialists, Patrick Waller, to form a team of experts representing the state of the art in pharmacovigilance. Patrick Waller is a medical doctor by training, who specialised early on in his career in drug safety. He worked in an isolated but nonetheless important research centre, the Drug Safety Research Unit, a research institute established by Professor William Inman, a precursor of pharmacoepidemiology in Great Britain and initiator of the UK system of spontaneous reporting. When called by his director to reflect on pharmacovigilance, Waller was then head of the MCA pharmacovigilance division and chair of the Pharmacovigilance Working Party at the EMA, but was considering reducing his intensive professional activities. At that time, the UK agency also employed a high-profile academic medical statistician. The executive director relieved them from other tasks for six months so that they could 'go away and basically think about pharmacovigilance and how it could be done better'. They worked with an in-house team involved in pharmacovigilance and also consulted colleagues globally by email. The team thus channelled widely accepted ideas emerging from drug safety experts in the WHO, FDA, academia and industry about overhauling pharmacovigilance in response to the challenges of complex internationalised systems of drug development and distribution. The result of that six-month brainstorming and synthesis was a report and academic paper putting forward a model for the conduct of pharmacovigilance (Waller and Evans, 2003).

The paper presented a critical evaluation of the effectiveness of pharmacovigilance as it was functioning then in terms of public health improvement: mortality and morbidity caused by unsafe medicines had not been

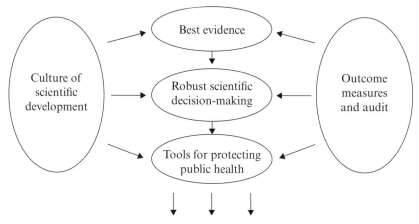

Figure 4.1   An excellence model for the conduct of pharmacovigilance (reproduced from Waller and Evans, 2003)

reduced in spite of increasing pharmacovigilance activities in the last decades. The paper acknowledged in a frank and honest manner that the development of pharmacovigilance had been deficient: it developed pragmatically over the years as a set of methods to detect the most serious reactions and avoid a tragedy like thalidomide, but with little concern for systematic regulatory and public health impact. Given the resources involved in drug development and the high public health and industrial stakes involved in dealing with adverse drug reactions – the detection of which can justify a revision of the marketing authorisation or its cancellation – its long-known limits (such as under-reporting of signals of adverse drug reactions) needed to be addressed. The authors observed that pharmacovigilance had to reach another level of reliability, precision or robustness to shape clear and credible regulatory decisions. The report exposed a hierarchy of types of data. It reasserted the now admitted idea that spontaneous reporting systems do not suffice any longer and other methods must be used alongside (Borden, 1981); controlled pharmacoepi-demiological studies in particular. It also emphasised the need to audit the system and its performance against measurable public health targets. Their article included an attempt to institute public health as a criterion of development of new tools and norms. They represented what they call an 'excellence model' for pharmacovigilance (Figure 4.1).

Measurable public health benefits are established as the primary quantitative criterion for the development and operations of pharmacovigilance.

It should inspire the acquisition of scientific methodologies, the classification of these methods and the data they produce.

Besides these programmatic elements, the paper put forward concrete propositions. The first of these in order of presentation in the paper is the concept of 'pharmacovigilance specification':

> A mechanism for achieving [the capacity to demonstrate safety rather than merely describe harm] would be for a 'pharmacovigilance specification' to be drawn up at the time of marketing authorization. This would explicitly consider the level of safety that had already been demonstrated, possible concerns that needed further investigation, and how further evidence was going to be gathered. It would be the responsibility of the applicant to draw up and maintain the specification, and it would be analogous to the Summary of Product Characteristics (SPC) in that it would require regulatory approval and become public at the time of authorization. (Waller and Evans, 2003: 19)

This is the first noted appearance of the term 'pharmacovigilance specification', which would eventually make its way into the ICH guideline for PVP and corresponding EU standards. As a regulatory concept, the proposition typifies a risk: the fact that by not knowing well in advance what is known about a drug and what is suspected not to be known, reactivity and the capacity to adjust regulatory responses to the appearance of adverse effects is diminished. As a correlate, the concept puts forward a concrete measure: requiring from manufacturers that they put together in one document this integrated assessment or image of the safety of their product, at the moment when it is introduced on the market.

The chain of events leading to the adoption of a standard based on this concept is straightforward. The ICH agenda was facilitated by the action of a British medical scientist, who I shall call Paul.[6] At this time, in 2002, he was an employee of the MCA and a member of the team set up by Waller and Evans, after several years spent 'running around hospitals' (in his own words). He was also the UK representative in the main European scientific committee, the CPMP, harboured by the European Medicines Agency. And on top of these various functions, he was representing the EU within the ICH medicine safety working group. He put the creation of a PVP guideline on the agenda of this working group, tabling a 'concept paper' in which the following idea is stated: 'carefully planned and effective pharmacovigilance activities, particularly for new drugs, can reduce the risk of drug toxicity and increase the benefit to public health. In addition, robust safety data can help avoid withdrawal of effective drugs from the market' (ICH, 2002). This concept paper, written by Paul with the support of other specialists, thus generalises even further the regulatory concept articulated within the MCA team. But it successfully led to the adoption of an international standard.

Lesson-drawing

The action of Paul to suggest the creation of a guideline on an idea emerging from the work of specialists is illustrative of the way in which lessons are drawn from the discovery of major adverse effects in pharmaceutical regulation. The PVP concept is the successful formulation of an idea that emerged around the year 2000 across the world of post-marketing drug safety specialists and out of a series of cases of high-profile adverse drug reactions. It is representative of this attitude of drug safety experts of lesson-drawing: reviewing the experience and knowledge of drug safety gained through series of cases, and assessing similarities and differences between them, to make deductions about ways of improving risk control. PVP supports the theory that there is much to be gained in speed and precision of remedial actions against unexpected adverse drug reactions, from better preparation and planning. Specialists made the same observation that insufficient sharing of drug safety knowledge between pre-marketing people and post-marketing people, within pharmaceutical companies and within regulatory agencies, meant a slower response to unexpected adverse drug reactions. Planning pharmacovigilance on the contrary would have the benefit of clarifying the state of knowledge about the drug's side-effects when it enters the market, as well as forcing pharmaceutical companies to more systematically provide information to regulators before critical adverse events appear.

Anticipation or planning is an idea that can be found in the proceedings of a conference of the International Society for Pharmacoepidemiology (ISPE, 2003). It was also part of FDA policy on the risk management of pharmaceuticals. The Japanese regulatory authorities were also at the time thinking around the concept of early-phase post-marketing vigilance. It was most clearly and decisively articulated in the academic paper published by Waller and Evans. The concept of planning crystallised through work of the MCA team, which channelled similar ideas that were being developed in disconnected places. Waller admits the idea of a pharmacovigilance specification was not the creation of a sole author:

> The world does not spin in a vacuum and, you know, things were going on elsewhere. The WHO, in some sense, was thinking in similar ways. And had their own project going. The FDA was thinking about risk management and had produced a concept paper. So there were quite a lot of similar things happening . . . which I think tells you something that a lot of people felt that pharmacovigilance had reached a point where somebody needed to go away, and think seriously to make the process better.[7]

The initiative of the MCA and the work of Waller and Evans funnelled an emergent common knowledge and gave it a definitive formulation.

With the publication of Waller and Evans' paper, PVP finally crystallised as a concept, backed by an academic reference. The paper now represents the source of an originally diffuse idea.

This repeats a historical pattern in the development of pharmacovigilance. The history of pharmacovigilance shows that its concepts were developed in this collective and inductive manner. Attention to high-profile events and subsequent research into systems of surveillance and methodologies of evaluation are constant (Routledge, 1998). Drug disasters always occur in the context of ongoing debates about the quality of existing regulatory technologies (Waller and Lee, 1999). For instance, thalidomide provided the catalyst for the reporting of suspicions of adverse drug reactions by medical doctors. It proved clinical trials to be inadequate for detecting rare adverse events because drugs are tested on limited populations, on specific profiles of patients, and during short periods of time. Soon after, it was realised that doctors here and there had suspicions of a phocomelia epidemic. Had there been a system for putting the thoughts of many doctors together, who may or may not see the link but may at least notice the health event, the cause of the event could be ascertained and corrected much sooner (Stephens and Brynner, 2001). The limitations of the spontaneous notification scheme were soon evident as new controversies about undetected or unmitigated events arose. In response, research focused on harmonising data collection (interconnection of national databases, formatting of data collection through forms and terminologies, creation of algorithms for causality assessment) and the development of different types of epidemiological studies with the objective of observing larger groups of patients or of prescription events (Berneker, 1992; van Boxtel, 1993). The concepts and tools of pharmacovigilance began to progress as new adverse events appeared that revealed what was previously unknown, pointing the way towards the development of new methodologies. The need to classify these methodologies and specify their benefits to study different sorts of adverse reactions, above and beyond the defence of a blanket system of notification and reporting, appeared greater than ever at the turn of the 1990s (Appel, 1998; Waller, 2001; Arlett et al., 2005). Some important issues in the history of the development of pharmacovigilance are shown in Table 4.1.

The concept of pharmacovigilance specification continues this process of drawing lessons from failures. One of the motives for promoting the idea of specification was the anticipation that, as the number of applications for new products slowed down between 2000 and 2005, a new wave of innovation was slowly mounting, for which adverse reaction issues would be different. This is what had happened with biotechnological medicinal products in the 1990s:

Table 4.1   Important events and issues in international pharmacovigilance (adapted from Lindquist, 2003)

Date	Product involved	Changes
1937	Elixir of sulfanilamide	Introduction of safety and quality checks in pharmaceutical regulation
1961	Thalidomide	National and international collection of adverse drug reaction reports
1970	Oral contraceptives	Realisation of under-reporting being a major problem with spontaneous reporting systems; Acceptance of the importance of epidemiological findings
1975	Practolol	Realisation that spontaneous notification cannot detect events not easily recognised as caused by a drug; Causality algorithms developed
1980	Non steroidal anti-inflammatory drugs	Development of pharmacoepidemiology
1982	Benoxaprofen	The USA saw the need to collect reports of adverse drug reactions that occurred worldwide; Start of harmonisation of data collection; Introduction of a causality algorithm in the legislation in France
1989	Fenoterol	Signals from case-control studies strongly debated; Use of database-nested studies become more accepted
1997	Third generation oral contraceptives	Focus on the need for good communications practice and consequence evaluation; Re-opens debate on issues of quality of evidence in pharmacovigilance

Drugs which are produced now, the way they are being produced, their target effects in the body, the way of being manufactured and so forth, may indicate that it is more difficult to predict their long-term effects. One example we've being dealing with is Remicade. Yes, it's given a lot of problem you know. The main problem with the opportunistic infection was undetected in clinical trials and emerged as a case report. That's partly because of its biological mechanism. You manipulate the immune system. Which is beneficial for the indication, but also it can be expected to create serious side-effects in patients with certain infections. So, that's just one example. All the biotech products which are being produced. I think we need a much better system than spontaneous reporting is, that could ever be.[8]

With pharmacovigilance specification, lessons are being drawn from the fact that waves of innovation tend to come with new types of safety issues that are poorly tackled by systems in place. This requires corresponding innovation in control. The work of Waller and Evans exemplifies this pattern of learning from series of cases and classes of events. Looking at the cerivastatin affair, both scientists centred their observations on particular classes of substances and mechanisms of action. It was at that level that they managed to typify risks and responses to them. Statins, for instance, have become interesting objects for the characteristic demonstration that clinical trials cannot ascertain safety in all dimensions. Even if the type of adverse reaction was known, it was the frequency of their appearance in larger populations that surprised everyone:

> Interviewee: Statins came as a series of substances, after the discovery that they were efficacious. But then one came up that did not want to behave like the others in the series. So the idea with risk management is that we know this family creates risks. So there is going to be a dynamic surveillance of all these risks for all of the products of the group. But that does not protect you from a surprise! So you stay on the lookout for new risks. Rhabdomyolysis was a known risk for the whole series. But cerivastatin did it more than the others.
>
> Interviewer: That's why it became such an important case?
>
> Interviewee: Well we got surprised by the frequency you know, it was something like ten times more, and with much more serious cases of rhabdomyolysis and hepatic problems. But all were posing problems since it was obligatory to monitor hepatic enzymes for all statins, at least in the first months after their marketing.[9]

Another standardised evaluation made by drug safety experts is regarding the lack of public health criteria to drive methodological and normative developments in pharmacovigilance. The early 2000s were a time of debate on pharmacovigilance, and in this context, the inductive approach towards methodological progress in pharmacovigilance is increasingly being questioned. Certain drug safety experts tend to minimise the ambitions of their science about designing evaluative methodologies. This is only an attempt to mask the difficulty in elevating it to the level of a systematic public policy towards drug safety. Queries about what drives methodological developments in pharmacovigilance arise. The director of the Uppsala Monitoring Centre (a technical centre funded by the WHO) thus argues that a shift from method-driven to public-health driven pharmacovigilance is necessary: 'The gaze of pharmacovigilance professionals has been preoccupied by gathering complete information and developing a more certain science. We must also be sure that individual patients benefit as much as possible from the information, they, the patients, give us' (Edwards, 1997: 86). Edwards' advocacy resonates with the call for a clearer definition of the objective in the development of new tools and norms in pharmacovigilance by Robert Nelson, a former FDA official. He argues that pharmacovigilance has always been developed in an ad hoc manner without clear understanding of the needs it is serving. The 'regulatory philosophy' of pharmacovigilance should be clarified (Nelson, 2000). In this sense, pharmacovigilance unfolded without any guiding criterion and with no other vision than the self-supporting search for better evidence. The measure of the overall benefits of introducing systems of collection and correction of adverse drug reactions has never driven the development of pharmacovigilance. It was rather seen as an adjunct to the individual exercise of therapy by the medical doctor. Paradoxically, the prevalence of injuries and fatalities induced by reactions to drugs is a rubric that is very much implicit in discussions about pharmacovigilance between its experts. The figures that exist are rarely mentioned. In their perspective, drug safety tends to be seen as a disaggregated and case-by-case activity. In other words, the prevalence of adverse drug reactions is not constituted as the overall public problem to which pharmacovigilance responds. It is this lack of any positive criterion that Waller and Evans try to fill by putting forward a public health argument in their paper. The emergence of this thinking about the development of pharmacovigilance shows that concepts such as pharmacovigilance planning were likely to quickly gain ground.

Overall, a striking element in the context of the appearance of PVP is the many allusions of drug safety experts to 'what they all knew' – of the risks of such a class of product, the interests of a particular fringe of the industry to avail of a new form of product surveillance, concern of the public for a given safety issue. One may thus conclude that experts had a common experience which, even if only partially articulated and shared, shows that they make the same assessment about what sort of regulatory practice to select and generalise. This experience directly serves to assess the changes needed in the practice and tools of risk assessment and management. Underpinning the whole discussion on PVP was non-public knowledge that such classes of products as Cox-2 and statins posed problems, and that these known safety issues were insufficiently considered at the marketing authorisation stage. To Royer, this pre-knowledge of safety issues, yet to be formalised by reports of adverse effects and epidemiological assessments, is what justifies developing risk management. It is effectively this expertise that legitimised the adoption of PVP. The overall discussion on risk management in drug safety was also informed by the perception that pharmacovigilance systems in place were inadequately efficient for new complex products, such as biotechnology-derived medicines. In 2005, in the midst of the discussion on the introduction of a risk-management strategy, Paul could thus argue that the delay in licensing applications that was then occurring was only the promise of another wave of innovation. This would lead to a new class of more complex adverse effects that surveillance systems had to be adapted to detect. Drug safety experts thus drew lessons repeatedly and formed the above typifications, and this was the basis for the emergence of a concept of pharmacovigilance specification and planning that they shared. Pharmaceutical risk management comes from a distinct source though.

Responding to Blame

The characterisation that is made of the statin issue, and the pharmacovigilance specification concept that responds to it, are thus best seen as lessons drawn collectively from a series of more or less politically salient events. The adoption of PVP and the impact of the concept on industry and regulatory agencies' processes are inseparable from the rise of a political risk-management agenda though, which was pushed in the aftermath of cerivastatin by EU regulators and politicians rather than by drug safety experts.

The cerivastatin affair was a point of divorce between regulators and politicians, with the latter suddenly blaming the system of pharmacovigilance and requesting remedial action. Members of the European Parliament sent several questions to the European Commissioner for Industry in charge of the regulation of pharmaceuticals. Questions were also raised by British members of parliament, while French ministers for health and social security asked the national medicines agencies to clarify the conditions for Bayer's decision and whether the withdrawal without prior information of regulatory bodies was lawful. European regulators were thus placed under political pressure to improve pharmacovigilance as a result of cerivastatin.

In the EU, the European Commissioner for Industry responded to questions from members of the European Parliament by asking the EMA for plans to improve post-marketing surveillance systems. Consequently, the EMA launched a 'risk-management strategy', summarising well-accepted ideas concerning the renovation and investment of more resources in pharmacovigilance. In parallel, the directors of national medicines agencies who were asked by their respective ministries of health and MPs to act, used their common platform – the so-called Heads of Agencies group[10] – to launch the development of their own risk-management strategy in the EU. Both strategies are similar and were merged soon after in 2008, with national agencies and the EMA forming a common committee for risk management. The strategies present measures for the development of an electronic database of adverse drug reactions, the reinforcement of quality assurance systems, transparency of their communication on the safety of medicines, coordination of resources for the conduct of pharmacoepidemiological studies and, lastly, exploration of methodologies in the conduct of pharmacovigilance.

The EMEA and HoA strategies show that the cerivastatin affair converted European regulators to the concept of risk management, following the FDA (Faden and Milne, 2008). The US regulator had established a risk-management framework in 1999, in response to a series of high-profile withdrawal decisions (FDA, 1999) that led Congress to launch a formal review of the FDA's action, on the grounds that such withdrawals may signify a too lenient pre-marketing evaluation, an unreactive post-marketing surveillance system or an inability to keep products on the market for specified, even narrow, groups of patients. The new FDA commissioner (who took office in 1998) consequently asked for a reanalysis of the withdrawals and for recommendations to be made on the improvement of pharmacovigilance. The exercise led to a report which denied that the reduction in the period of pre-market regulatory review of drugs has led to more withdrawals (or to the licensing of more unsafe drugs). The report thus validates the slow policy shift towards reduced pre-market review (Olson, 2002; Ceccoli, 2003; Abraham and Davis, 2005), by arguing for the development of a more effective post-marketing surveillance system to compensate for quicker pre-market assessment. The report expanded on the complexity of the marketing system and use of medicines. The argument is that the safety of medicines can no longer be ensured through doctors' prescriptions and notifications. A number of factors cause medicines to be used in the wrong way and generate adverse effects – such as the increasing number of life-style medicines, self-medication, distribution of products via the Internet or without prescription, increasing numbers of so-called me-too molecules, and so on. The traditional chain of controls that extends between the development of the product and the patient's intake (pre-marketing evaluation, prescription by the doctor, delivery by pharmacist, collection of signals of adverse drug reactions) no longer suffices. In this context, the arsenal of risk minimisation measures must be extended beyond 'dear doctor letters',[11] variation decisions[12] and product withdrawals. The concept of risk management emphasises that the activity of controlling safety neither starts nor ceases once the product is on the market (Andrews and Dombeck, 2004). It is a continuous trail of actions that extends all along the product's life cycle. This led the FDA to programme a series of reforms on the measure of risk before marketing, the use of observational

pharmacoepidemiological studies and the spread of 'good pharmacovigilance practices' in the industry.

Emulating the FDA and faced with similar circumstances, European
regulatory agencies arrived at the lessons of the cerivastatin affair
through their own risk-management strategies. The rationale of the
European risk-management strategy of national agencies is explicit on this
lesson-drawing:

> European systems for monitoring the safety of medicines in clinical use and
> taking action to minimise risk have come under scrutiny, particularly following
> a high profile drug withdrawal in 2001 [cerivastatin]. A review of the lessons
> learnt in Europe from experience in managing drug safety issues was con-
> ducted. Based on casework history rather than on objective outcome measures,
> its extent and diversity nonetheless fully justified the preparation of a coherent
> European Risk Management Strategy. The process of bringing together exam-
> ples of lessons learnt also illustrated the need for core performance standards,
> an effective issues tracking and management system, and agreed methodologies
> for audit and monitoring outcomes. (HoA, 2003: 3)

The initiatives of the FDA, EMA and HoA draw from the same ideas,
such as the necessary coordination of regulatory agencies, better use of evidence and safety information, and the development of risk communication.
The introduction of pharmaceutical risk management by regulators cannot
be dissociated from several broad policy priorities: the increased coordination of regulatory decisions of separate national agencies; the investment
of more public resources into the running and management of information
systems, including the European IT system of adverse drug reaction collection called EudraVigilance; and the quality of regulatory decision-making.
The heads of national regulatory agencies, leaders of the EMA and the
European Commission share these objectives (EMEA and HMA, 2005a,
2005b and 2007). These objectives, which are inseparable from a form of
blame-avoidance action, were even more critical after September 2004 and
the highly controversial withdrawal of Vioxx medication.

Vioxx (non-proprietary name Rofecoxib) is an anti-arthritis non-
steroidal anti-inflammatory drug of the 'Cox2-inhibitors' class. Cox2
drugs have been controversial ever since their marketing, because they
were claimed to provide additional benefits to existing anti-inflammatory
drugs, but also to bring in new risks. The risk/added benefit balance was
judged sufficiently positive to allow the drugs on the market at the end
of the 1990s. Vioxx is estimated to have been prescribed to 80 million
people worldwide and has generated billions of dollars of revenue for
its manufacturers Merck. In 2004, an ongoing clinical trial revealed that
Vioxx doubled the risks of stroke and heart attack at high dosage and in

long-term use. Merck withdrew the product from all national markets at the end of September 2004. The media fall-out around the withdrawal was probably larger than in the case of cerivastatin, given the number of patients affected and the number of suspected deaths attributed to the drug, as well as the withholding by the manufacturer over several years of data demonstrating this increased risk. In terms of regulatory reform, however, this case has only emphasised the need for the measures that had already been articulated in the aftermath of cerivastatin.

The European risk-management strategies did include a reference to PVP. The adoption of a guideline is one of the measures in the overall plan for modernisation of pharmacovigilance in Europe. It was, however, no longer mentioned in the European risk-management strategy after the ICH work on the PVP guideline was launched. The spread of the maxim of pharmaceutical risk management thus explains the ease with which the concept of pharmacovigilance planning was endorsed in the ICH, where delegates of all three major regulatory bodies are represented, along with industry associations. The political consensus around risk management thus accelerated the adoption of a PVP guideline and the endorsement of a slowly emerging and collective concept of pharmacovigilance specification.

This situation did not necessarily please all drug safety experts, more inclined to develop specific new techniques with a proven impact on public health – as PVP should be – rather than deploying all-encompassing programmes of regulatory reform under the heading of risk management. In the approach of analysts studying adverse drug reactions, pharmacovigilance planning is here to facilitate the explication of everything that is known, to better define what remains to be known (in this particular case, rhabdomyolysis was known, but not the rate at which it appears in the overall population). In their mind, pharmacovigilance planning is a tool to work on such specific safety issues, not a blanket concept to demonstrate the relevance of pharmacovigilance. Risk management, on the contrary, is to them a political response to an affair such as that of cerivastatin, which caused political fall-out but taken on its own did not carry any new lesson on the conduct of pharmacovigilance.

> Politicians realised that pharmacovigilance was important with Lipobay [Cerivastatin]. Pharmacovigilance is a strange word. But in 'risk management' you have risk and you have management. It was like there was a need for a new term. It is a new word for new money. Insider people know what we want to say, which is strengthening pharmacovigilance.[13]

In other words, risk management is the result of politicians' intervention in a topic that was new to them, but cerivastatin was not a special case to drug safety experts, only a classic example of the limitation of clinical trials. This instrument cannot inform about all safety issues of the product before marketing. The case only confirmed that pharmacovigilance is an inherently limited system since suspicions of adverse drug reactions are under-reported and the data collected imperfect. However, this was known already. They hypothesised that the distinct lesson of cerivastatin is that it remains difficult to measure the relative safety of products in the same class (all statins cause rhabdomyolysis, albeit not at the same frequency) and to treat these products accordingly. The measure of these variations and the subsequent adjustment of marketing authorisations require sophisticated instruments and years of data collection. Regulatory agencies do not have the means to act quickly and with precision. The discourse of regulators on risk management gives little space to this notion of methodological progress and specific lessons. Risk management seems to be much more about balancing different types of risks (risks to the industry of withdrawing a product, risks to regulatory agencies' reputation when not doing so, risks to the patients) than about demonstrated progress in public health (Ceccoli, 2003), as drug safety experts would like it to be.

Another expression of discontent by some drug safety experts is the opinion that this quick process of harmonisation is incompatible with proper methodological experimentation and the need to assure that methodological changes are of benefit to public health. The prospect of an internationally adopted rule that responds to recent regulatory failures provided the reason for shortening the time in experimenting with this new methodology. The introduction in 1995 of the Periodic Safety Update Reports (PSUR) is a similar example in which a promising, but not fully tested tool was rapidly adopted worldwide (Castle, 1992). Manufacturers must submit these reports at regular intervals to present all safety warnings received in previous years and build an overall image of the product's safety. The ICH adopted a guideline for this tool following its elaboration by CIOMS.[14] This committee was used by a group of pharmaceutical manufacturers as a policy venue, to reflect along with scientists and independent medical doctors on concepts of pharmacovigilance. It took a good number of meetings to develop a satisfactory version of the concept and an accepted name for it. It then took 15 years for PSUR to be refined and another five years for the concept to be integrated in European law. The PSUR tool has been criticised in the aftermath of cerivastatin. PSUR requires setting up information infrastructures, hiring specialists, and devoting enormous amounts of time to collecting, compiling, evaluating and reporting information. A PSUR is typically several hundred pages thick and takes time to review. It fails to provide clear criteria for regulatory decisions and its impact on the overall mortality of drug-induced disease is doubtful. The introduction of PVP is yet another case of innovation in safety process and tools, driven by an opportunity to adopt and implement consensual concepts. Only two years elapsed between the formulation of the proposition in Waller and Evans' paper and the adoption of the ICH guideline. The same effort and resources for PSUR were needed to turn the PVP idea into routine practice in companies and agencies. When the creation of a guideline for PVP was put on the ICH agenda, none of the public health benefits of these enormous resources and new practices could be established. For Waller, for instance, these resources could just as well be allocated to the implementation of other tools (such as multi-centre pharmacoepidemiological observational studies) with more profits for public health.

These arguments express the dissatisfaction of drug safety experts, more inclined to methodological progress and public health achievements than to political blame-deflection and regulatory harmonisation. This did not prevent standardisation happening: there has been a junction between the circuit of regulatory lesson-drawing that the team of Waller and Evans mobilised and enterprises of international harmonisation and standard-setting organised around the ICH. This happened through a particular fringe of drug safety experts and medical scientists turned quasi-regulators. The next section describes at greater length the action of Paul and of another scientist, Gabriel, who represent this smaller college that was decisive in standardising the concept of PVP to form an international and European guideline.

PHYSICIANS, DRUG SAFETY SCIENTISTS AND REGULATORS

The PVP guideline produced in the ICH and the risk-management guidance produced by the EMEA following this illustrate that research in pharmacovigilance methodologies and global regulatory harmonisation has been further integrated, which is a side-effect of the circulation of drug safety experts across regulatory bodies, such as national and European regulatory agencies as well as the international standard-setting arena of the ICH.

The original concept of pharmacovigilance specification emerged from an open set of drug safety experts who worked to typify its problems and possible improvements. This set of geographically distant experts is shown in Waller and Evans' paper. At the end of this paper, the two authors thank 58 people for sharing thoughts and ideas with them. And indeed,

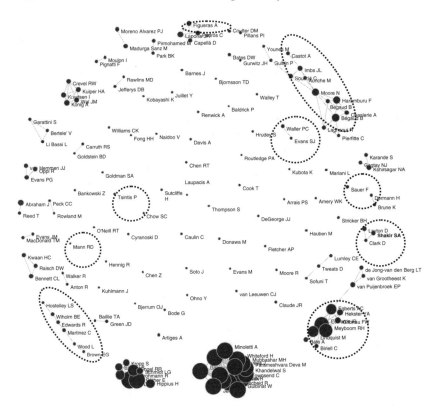

Figure 4.2   Map of co-authorship networks in the field of
pharmacovigilance planning

the authors' work was based on a collegial exchange of ideas within the small world of drug safety experts. They consulted leading figures of pharmacovigilance across these groups, in conferences and through email exchanges. Many of them are visible in Figure 4.2, illustrating the most prolific co-authorship groups in pharmacovigilance.[15]

The map in Figure 4.2 represents the most published authors in PVP research and related themes as well as their collaborations. It illustrates the extent of fragmentation in research on pharmacovigilance. There are indeed many authors in this field (which numbers around 500 publications in total), demonstrating the degree to which publications are co-authored. It also denotes that the field is well established, that the competition to publish is strong and that publication in academic journals is an important and accepted vector to discuss mixed scientific and regulatory issues as pharmacovigilance typically is.

Authors who are clustered are those who collaborated through publications. They often belong to the same research institutes, university hospitals or regulatory agencies. They are also from the same country. Figure 4.2 shows that research (through the proxy of scientific publications) on an innovative topic like PVP is done simultaneously in a great number of research centres which are formally independent. Tellingly however, a good number of people contacted by Waller and Evans are present on this map (circled in Figure 4.2). These contacts have created an informal and provisional college of reflection on pharmacovigilance (informal in the sense that it did not result in academic publications, since there is no linkage between these people or groups on the map). All of them are academics or researchers in drug safety with a hands-on position in pharmacovigilance. For instance, the upper right cluster is the French school of pharmacovigilance, including Bernard Bégaud, a university hospital professor of pharmacology and one of the leading figures of the French school of pharmacovigilance, as well as Anne Castot, then head of the EMEA pharmacovigilance working party and director of risk management in the French medicine agency. The bottom left group includes the scientists of the World Health Organization international drug monitoring centre (so-called 'Uppsala Monitoring Centre'). The bottom right group represents the Swedish school. This figure, together with the more qualitative information concerning the consultation work of the MCA team, explains why the various reports on risk management contain the same ideas, such as that of PVP. They were put forward by a large set of specialists, attached to those regulatory agencies, in a position to influence the field by their capacity to publish in it and informally engaged in an circuit of standardisation that was mobilised by the team of Waller and Evans.

This circuit had thus already produced a concept of pharmacovigilance specification that was widespread, not only in academia but also among regulatory agencies, given the fact that so many of the above experts themselves work in regulatory agencies or in their scientific committees. This explains why the introduction of PVP was an accepted fact in CIOMS and ICH, as recounted by Paul:

> Paul: So I took this part of the excellence model and took it in the international arena.
> Interviewer: And what was the reaction of the people there?
> Paul: Erm, it was opportunism. Because I was representing the EU in ICH, for the whole of pharmacovigilance. And we were brainstorming what new topics we might produce guidelines on. And because I was quite evangelical about this particular topic, I pushed it through on behalf of the EU.
> Interviewer: And were people prepared?

> Paul: Yes, because although it was novel, and is novel, a lot of people had similar thoughts at the same time, that there was not enough planning in pharmacovigilance basically. So when somebody like me comes to the international arena and says 'let's have a guideline that gives a structure and a format in which these plans must be written', which is what essentially this guideline is, I was pushing an open door.[16]

Paul thus admits that the ICH agenda was formed on the basis of an already well-accepted concept that planning would bring about positive effects in the management of drug safety and should be pushed by establishing rules for it.

Paul thus forms, by his circulation and opportunism, an important connection between drug safety experts and international regulatory standard setting. A second point of connection is constituted by Gabriel, a French doctor and drug safety expert, who supported the initiative of Paul and coordinated with him, thus embodying a smaller invisible college that is in great part responsible for the standardisation of PVP. Gabriel is the qualified responsible person for pharmacovigilance in a multinational pharmaceutical company, alongside working as a general physician. During his career, he developed a particular interest in drug safety and in methodologies for the monitoring of adverse effects, in clinical trials and post-marketing. He is on a European pharmaceutical industry committee on regulatory affairs and a regular participant in meetings of the Council for the International Organization of Medical Sciences (CIOMS). The CIOMS is a WHO-attached but autonomously funded professional body that initiated a harmonisation of terminology and practices of pharmacovigilance in the mid-1980s (among other international medical standards). On pharmacovigilance matters, it has been used by the pharmaceutical industry, especially by companies most advanced in pharmacovigilance processes, to work with the medical profession on international standards. Since the ICH was founded in the early 1990s and progressively became the central standard-setting body in pharmaceutical regulation, the CIOMS has been used for sort of pre-ICH meeting expert consultations. Gabriel also sits as the industry representative in the ICH working group on medicine safety, along with Paul.

Gabriel had worked with colleagues on a 'Development Safety Update Report' (DSUR), which is similar to PVP. The seventh CIOMS working group thus spent a series of meetings reflecting on the content and format of such a dynamic document, which traces the evolution of the knowledge of the safety of the product as it progresses through stages of development including toxicity and clinical testing, licensing and pharmacovigilance. This idea of a DSUR follows from a previous innovation of CIOMS, which was adopted worldwide as the PSUR – a recapitulation of all safety events that manufacturers have to provide to regulators at regular times. The template for the DSUR developed by the CIOMS contains rubrics that resemble those adopted in the ICH, such as Update on Actions Taken for Safety Reasons, Inventory and Status of Ongoing and Completed Interventional Clinical Trials, Information from Marketing Experience, Summary of Important Risks (CIOMS, 2006).

On several occasions, the CIOMS has been used as a forum where guidelines later developed in the ICH were first tested. This is, again, what happened in this case. Gabriel conducted the work within the CIOMS, in cooperation with Paul. They introduced the topic in the ICH together, as members of the ICH safety working group. For these two experts, the objective of creating a guideline was programmatic and organisational. The plan of Paul and Gabriel in going to the ICH was not so much to impose these changes using a mandatory instrument, but to create the expectation in a whole system of intermediaries that risk management will happen – companies that provide specialised services in patients' surveys and database management, training and consulting businesses in risk management, and so on – and to spark a series of reorganisations in companies and regulatory agencies. As argued by Gabriel, before the adoption of the ICH guideline:

> The minimisation of risks for the patient is going to be, in a way, the trigger of an array of processes within pharmaceutical businesses, as well as of specialisations of people and finally of reflections by various people in industry and elsewhere around the methodologies and procedures to put in place to anticipate and measure risks.[17]

They sought to make companies more sensitive to the idea of anticipation and to offer a technology that would support the reorganisation of companies. The guideline provides ways to construct procedures and informational devices that break the barriers between pre- and post-marketing safety people. In other words, their ambition is to have the industry as a whole progress in its post-marketing surveillance practice and evolve towards those planning practices that as practitioners and students of pharmacovigilance they see as most effective. Their ambition did not go as far as reshaping the regulation of medicines and channelling more public health resources into a new tool that would revolutionise drug safety surveillance. More modestly, they sought to raise awareness within companies for the betterment of planning practices. As illustrated above, the ICH guideline on pharmacovigilance planning was absorbed into EU guidance as 'risk-management' plans. Specialists' and regulators' concepts have both been absorbed competitively into the EU standard. This can be explained by the fact that the invisible college behind the initial idea of pharmacovigilance planning is embedded in regulation.

Fifteen years after its first participation in the CIOMS, Gabriel tellingly presents CIOMS-related activities to a portfolio of other more 'statutory' ones: representing his company in an industry working group and in the ICH. His participation in CIOMS work has become closely integrated with his action in the ICH. In his own words:

> I am part of a certain number of associations or technical committees that reflect on the scientific dimension of pharmacovigilance. It is the interesting aspect of this expert activity to ensure the progress of pharmacovigilance as a discipline. But we also reflect on the regulatory aspects as well, because pharmacovigilance is a highly regulated activity. And for a good reason: we, the industry, were the first to ask governments to introduce regulations [. . .] The apex of regulation, it's ICH. Because that's where we managed to enact the concepts which we developed in CIOMS. The CIOMS is like a think tank, a reservoir of ideas for ICH. Within ICH, the same people that meet in CIOMS as independent experts meet as industry or government representatives.[18]

Gabriel thus expresses the fact that as the EU adopted a legal framework for the authorisation of medicinal products and post-marketing surveillance, and as international harmonisation accelerated through the ICH, the development of the science of pharmacovigilance became intertwined with the production of corresponding norms. On the point of innovating through harmonisation:

> The other thing that is very interesting at ICH, which is something I have been involved in so I have a direct experience, is that ICH is about harmonisation. So what has happened in almost all cases so far, at least one of the parties, or two or three, has already got guidelines, and they come along and they fight out a harmonised guideline, which is basically what everybody can sign up to. A compromise. What they're doing here with PVP is different. They are not harmonising; they are actually building something within a process that is designed for harmonisation. That may be possible but that may be slightly dangerous, it's a change. ICH is now starting to build something from scratch, and that needs to be recognised. It needs to be thought about, actually, whether it is the right place to do it.[19]

In this process, older medical scientists and professors who initiated pharmacovigilance in close cooperation with their national ministries and the WHO, were overtaken. An almost self-nurtured regulatory system now produces norms much more quickly than those scientists could hope to do. Some indeed were annoyed that the ideas explored in the CIOMS had been 'stolen' by the ICH. A related observation by drug safety experts, particularly those involved before the Europeanisation and internationalisation of medicine regulation (creation of the EMEA and of acceleration of the ICH as a result of the nascent collaboration between the former and the FDA), is that academics' ideas are increasingly 'pilfered' for consideration within regulatory arenas such as the ICH. To Royer, one of the initiators of the French pharmacovigilance system, founder of the European Society of Pharmacovigilance and active participant in the pre-1993 meetings of the European Pharmacovigilance Working Party, the CIOMS functioned ideally in gathering together anyone willing to contribute to advancing pharmacovigilance, from academic pharma-coepidemiologists to industry safety officers, with the view to make sure data were circulating. The ICH is on the contrary designed to turn out standards, and has been amazingly effective in doing so to the point of having to search for topics to produce standards on after three years of functioning. They also regret that scientific learned societies have been superseded in the orchestration of the reflection on pharmacovigilance by professional conference organisers and drug information companies. These close connections between academic research and standard-setting accelerate the standardisation of concepts like PVP and risk management and implementation, above and beyond experimentation and research on them. When Waller regrets that PVP was generalised too quickly, he only underlines this broader structural trend that some of his colleagues such as Paul and Gabriel contribute to.

In contrast to their peers, Gabriel and Paul have integrated the need to act with regulators and businesses rather than within the confines of their professional community. In this new setting, medical scientists are influential only if they become polyvalent actors, mixing their technical expertise with norm-setting aims and strategies. Paul's trajectory and action indeed illustrate this. In 2003, he became a European Commission official and a policy-maker. He worked on the adoption of several pieces of European law before becoming head of pharmacovigilance at the EMA in 2008. This cadre of regulatory experts emerged from the crea-tion of regulatory agencies which formed its natural habitat. Agencies harbour committees that medical scientists sit on, sometimes very early in their career. As they take on a role of expert at the national level, they are often drawn into a similar function at the European and international level, thus beating out a career path that is much more regulatory than scientific.

The views of Paul and Gabriel on the risk-management agenda of regulators are also symptomatic of the work they did to align with regula-tors' ideas. According to Paul, risk management adds very little to what pharmacovigilance practitioners knew and did. It is a concept introduced opportunistically by a handful of officials in European and national agencies in response to political fallout. They felt the latter needed to deflect blame, boast about the decisiveness of imaginary changes in response to the affair, and advertise their capacity to steer still rather recent agencies. But the real point in risk management, and the correct expression to use, is that of 'risk minimisation'. What is specific to this context in which drugs are frequently withdrawn is the need for better techniques to minimise risks effectively, by impacting rapidly on prescriptions and consumption behaviours, other than by merely withdrawing a product. Cerivastatin revived the view that tools for anticipating or acting quickly and effectively on the level of risk through communication with patients and doctors are under-developed (Lindquist, 2003), something that the PVP guideline starts remedying. To Gabriel, even more strikingly, risk management is enfolded in what pharmacovigilance has always pledged to do:

> Of course we must work consistently from beginning to end, on the same product and with the same person. Pre- and post-marketing safety are not two different jobs, this is a matter of course. The management of the risk is done in the same way throughout. It is just that there are tools you can use in the pre-marketing time, during clinical trials, because there is control over the denominator, the exposed population. Whereas it is slightly different in the post-marketing domain, in which we proceed through statistical approximations. But all of these are tools . . . Risk management is properly understood pharmacovigilance, modern, dynamic pharmacovigilance. Risk management is what anyone does in pharmacovigilance when it's done correctly. It is just that it has now been better materialised, better conceptualised. It is pharmacovigilance writ large.[20]

According to this discourse, risk management is the rejoining of the clinical evaluation of safety and of pharmacovigilance and pharmacoepidemiology under a new terminology (Hyslop, 2002). It simply emphasises the importance of reconciling pre- and post-marketing safety, but does not invent this imperative. For older proponents of pharmacovigilance, risk management, as a dynamic drug safety assurance that combines pre- and post-marketing activities, has been the fundamental project of pharmacovigilance since the beginning. This is particularly true for the person interviewed here, who introduced his company to the practice of systematically collecting and assessing adverse drug reactions signalled during clinical trials, a decade before it became mandatory.

These two were thus central, by virtue of their position and ideas, in connecting different regulatory actors. In relational terms, they connected the set of medical scientists and professionals from whom the concept of planning emerged, to the international standard-setting arena of the ICH and to regulators. Their relation with regulatory agencies and the ICH embedded a college of pharmacovigilance practitioners and researchers in regulatory organizations, to constitute the 'small world of pharmacovigilance' characterised by Paul: 'Pharmacovigilance on the one hand is a big world because there are quite a lot of people employed globally, because of all the requirements. Particularly on the industry side. But there is a much smaller number of people that gets involved at a senior level'.[21] Among those people stand medical scientists like Paul and Gabriel, whose circulation was decisive in linking the evaluative or lesson-drawing activity of experts and the standard-setting drive of regulatory agencies. This integration explains why, in spite of a greater awareness in the community of pharmacovigilance of the necessity to experiment with methodologies to ascertain their public health benefits, a new pharmacovigilance concept was so quickly considered in regulatory arenas and put into effect.

Paul and Gabriel are not exceptional cases however. Other members of the MCA team took part in the meetings set up by the HoA to design the risk-management strategy. The risk-management strategy launch report explicitly builds on the ideas of the MCA team and indeed cites Waller and Evans' paper as a result. Waller himself is an interesting illustration of a link between research and regulation, by means of which more numerous points of contact have appeared, and it is through these that concepts can be translated into standards. Originally a drug safety scientist, he joined the MCA in the early 1990s and then rose up the hierarchy, becoming chairman of the EU Pharmacovigilance Working Party by the end of his career. Further to this, the majority of the 58 people who Waller and Evans thank in their papers are linked to a regulatory body either as their main activity or as a side-activity.

As hinted in Chapter 2, clinical pharmacologists, often with a foot in hospitals and medical practice, have been instrumental in the early development of pharmacovigilance, simultaneously in their countries and in the WHO. However, in the case of PVP and of risk management more broadly, they form a productive set of relationships that turns out new methodologies and norms much more quickly than before, as some of its members are active in transnational regulatory institutions such as the EMEA, the European Commission or the ICH. A close study of the profile of the PVP proponents helps to illustrate these two aspects: the growing competition between medical scientists turned quasi-regulators – they make concepts for rules and rules themselves – and the declining influence of national medical schools in an increasingly transnational normative production. This becomes clear when we follow the medical scientists proposing the PVP concept. They navigate between the big players – the three major regulators, large pharmaceutical companies

and their industry associations, as well as dominant professional associations. Ideas that are agreed between these entities find themselves
rapidly translated into formal guidelines, bypassing the experimental
stage of testing these new practices and techniques. PVP is the result of
a system shaped by scientists acting as quasi-regulators rather than as
researchers.

The dissatisfaction with the rapid shift to enforcement of an experimental idea of planning is a symptom that the centre of the regulatory
system has moved: from hospital practitioners and the WHO or CIOMS,
to regulatory agencies and the ICH. Waller and Evans' team produced an
already quite standardised form of PVP, with a conceptually clear idea of
what its content should be. But it was Paul who made the definitive act
of standardisation. Waller was not far from reproaching his younger colleague for having tried to achieve fame in his name at the ICH by investing
in an appealing idea. Sociologically, the individual trajectories of Gabriel
and Paul illustrate the fact that pharmacovigilance has become both international and regulatory in nature.

The system transforms the product of academic-style exchange into
formal standards much more rapidly than in the circumstances that
drove the setting up of national spontaneous reporting systems in the
1960s and 1970s, and which took years to institutionalise as regulatory
systems. It is, fundamentally, the same college of scientists who are at
work, except that the trajectory of some members towards functions with
a greater regulatory dimension has changed the nature of its products.
Reflecting on the science of pharmacovigilance sees it intertwined with
producing 'regulating' effects, which explains the imperfect alignment
on the meaning of risk management and on the public health impact of
the introduction of PVP. This is explained by the shape of an invisible
college, of which Paul and Gabriel here have been part of a cluster made
more dense by the presence of increasingly numerous regulatory agencies
(for which standards and norms are an essential instrument) and by an
overarching international organisation (ICH), where a few drug safety
experts convey the ideas of researchers and medical professionals to the
regulatory world.

CONCLUSION

PVP is the story of the standardisation of a concept of planning and
illustrates strikingly the close connection between the evaluative work performed in circuits of drug safety specialists and processes of international
harmonisation. This junction was made in this case by a small college of

drug safety experts with multiple official functions in regulatory agencies, in the industry, as well as in transnational bodies like the CIOMS and ICH. By articulating a more unified concept of pharmacovigilance specification and planning, they made it easier to create a harmonised international guideline. They also aligned practices of pharmacovigilance with a political risk-management agenda.

The junction between evaluation and standard-setting was not seamless though. Differences of opinion between scientists concerning the benefits of harmonisation processes reflect the internationalisation of standard-setting, the acceleration of regulatory-driven harmonisation, and partly its distance from lesson-drawing and experimentation of new practices. Through being negotiated by regulatory agencies and industry associations, the implementation of these standards is much more imperative, and their inclusion in national or regional bodies of law more mechanical. As a result, drug safety practices are standardised through formal process of norm adoption, rather than through experimentation and diffusion throughout the industry or the medical profession under a less formal status of best practice. Standard-setting is more formal, with participants as representatives of regional regulatory authorities or of business associations, rather than experts participating in their own name. Scientists involved then seem to show more awareness of the political (power plays between regulatory agencies) and legal aspects (the adoption of an ICH standard is a direct prescription of new practice imposed on industries) than their predecessors. The PVP case among others, and the distinct trajectory of Gabriel and Paul, illustrate that regulatory intervention in drug safety changes under the influence of more formal, transnational and rule-oriented forms of medical expertise. Medical experts in that context remain highly influential and central, as they circulate among other actors, hybridise their professional expertise with elements of legal and managerial knowledge, and demonstrate good skills of strategy and negotiation, all of which appeared critical in the successful promotion of PVP in the ICH, its integration in a risk-management framework and the alignment of all regulatory actors. Thus, the standardisation of PVP represents a sort of pacification of the relations between opposed medical schools of pre-marketing clinical testing and post-marketing surveillance, all being outpaced by those medical scientists who have a foot in standard-setting arenas. All of this is the indirect consequence of the internationalisation of regulation and the increased recourse to impersonal expert-based rules expressing generic practices, which can be agreed upon by multiple regulatory bodies and transferred across geographical frontiers.

NOTES

1. The ICH was launched in the early 1990s with the objective of harmonising regulatory requirements of the three major pharmaceutical regulators: the FDA (US Food and Drug Administration), the European Union and the Japanese Ministry of Health. Representatives of the FDA, Japanese authorities, European Union and major trade associations participate in its meetings. To date, it has produced 69 guidelines (see www.ich.org/LOB/media/MEDIA356.pdf, consulted on 4 January 2010). ICH represents a rather extraordinary case of globalisation, in a sector in which governments long held on to their autonomy. The ICH success (Vogel, 1998) cannot be understood outside the Europeanisation of pharmaceutical regulation: through the early 1990s reforms, national European governments have become aware of their interdependence and of the benefits of cooperating. The EU has been very active in ICH and tends to carry out its guidelines quite faithfully.

2. These are Redux, Pondimin, Seldane, Duract, and Posicor in the USA (see FDA, 1999; Friedman et al., 1999) and Baycol/Lipobay and Vioxx globally respectively in 2001 and 2004.

3. This was the motivation for reorganising the UK Medicines Control Agency (MCA) at the end of the 1990s along those lines, instead of product teams in charge of a particular group of products.

4. Commercial names included Staltor, Cholstat, Lipobay, Baycol.

5. In the European regulatory system, the licensing of certain categories of products is collegial: a scientific committee comprising national experts collectively prepares decisions. The scientific evaluation of products is entrusted to one national expert, who acts as rapporteur for the rest of the scientific committee. This expert or, in practice, the agency he/she is the employee of, remains in charge of the product after authorisation.

6. In the rest of the book, I use pseudonyms to speak about the scientific experts who I take to be part of an invisible college.

7. Interview with Patrick Waller, 5 February 2004 in Southampton.

8. Interview with Ingmar Persson, 8 October 2003 in Uppsala.

9. Interview with René-Jean Royer, 16 January 2004 in Brussels.

10. HoA is the acronym for Heads of Agencies, first an informal reunion of the directors of national medicines agencies of Europe, who developed the habit to meet and discuss issues of common interest in the wake of the creation of the EMEA. The HoA has no official role, but seizes issues of importance whenever it feels it can contribute to improve policy coordination (or, in less diplomatic terms, cordon off the growth of European regulatory structures). It is now called HMA, for Heads of Medicines Agencies.

11. This is the standard name for official written communications addressed by regulatory agencies to physicians concerning changes to the permitted uses of a drug.

12. Adjustment of the terms of the product licence and its Summary of Product Characteristics (the list of specified properties and adverse reactions of the product, which legally defines the marketing claims that may be used by the manufacturer and the content of the product's usage notice).

13. Interview with Priya Bahri, 6 February 2004 in London.

14. The Council for the International Organization of Medical Sciences is a nongovernmental organization jointly supported by UNESCO and the WHO. It gathers representatives from professional and scientific bodies of the medical world to establish standards related to such topics as medical ethics, use of medicines and medical terminologies.

15. To produce these maps, a search through keywords describing the regulatory concepts was made in a major database of scientific publications (Medline for PVP and PMM, Web of Science for HACCP). Réseau-Lu then represents on a map the authors who have published the greatest number of papers (who, according to the quantitative approach of Diana Crane, thus form part of the 'invisible college' of the field, as

opposed to larger group of collaborators). The settings chosen to produce these maps are the following: the 30% most widely published authors in the field are selected, and the size of the point by which they are represented is proportional to the number of publications. The links represent the most frequent collaborations between these authors, which the software cluster together to make them more visible. Authors that stand alone are the ones who publish a lot either alone or with co-authors who do not publish as frequently.

16. Interview with Paul, 30 January 2004 in Brussels.
17. Interview with Gabriel, 27 January 2004 in Paris.
18. Interview with Gabriel, 27 January 2004 in Paris.
19. Interview with Patrick Waller, 5 February 2004 in Southampton.
20. Interview with Gabriel, 27 January 2004 in Paris.
21. Interview with Paul, 30 January 2004 in Brussels.

5. Modelling regulation: HACCP and the ambitions of the food microbiology elite

In 2003, the EU adopted a regulation that mandated all food businesses to have a Hazard Analysis Critical Control Point (HACCP) system in place from the year 2006. This legislation extended the obligation already set by Directive 93/43 to organise food hygiene by following HACCP-like principles. The Codex Alimentarius Commission, an entity supervised by the World Health Organization and Food and Agriculture Organization, had also adopted a guideline for HACCP in 1993. These norm-making acts capped the long course of the concept of HACCP that emerged at the turn of the 1970s in reflections involving NASA, a US Army laboratory, the US food company Pillsbury[1] as well as the US Food and Drug Administration before becoming a global and generic methodology for food hygiene and safety control.

HACCP is a system of operations to ensure that contaminations of a foodstuff under production are minimised. The principles are as follows (Codex, 1997):

- Conduct a hazard analysis.
- Determine the critical control points (CCPs).
- Establish critical limit(s).
- Establish a system to monitor control of the CCP.
- Establish the corrective action to be taken when monitoring indicates that a particular CCP is not under control.
- Establish procedures for verification to confirm that the HACCP system is working effectively.
- Establish documentation concerning all procedures and records appropriate to these principles and their application.

The HACCP acronym was forged to group together these operations that must be followed in a set order. The application of these principles leads to the elaboration of an 'HACCP plan' for monitoring and correction of potential incidents by companies. In such a plan, the production process is represented as well as the critical points along the chain, and the threshold of presence of the different substances to monitor. Typically, an 'HACCP officer' is tasked with setting up and monitoring the execution of the plan, leading a team of people with competences in food analysis and food hygiene as well as risk assessment to do so. The HACCP officer also liaises with official inspectors to demonstrate the diligence and overall level of safety maintained in the production. This form of continuous monitoring has been conceptualised to replace costly and ineffective end-of-chain testing, in a way that announced the contemporary rise of performance- and process-based food safety regulation (Henson and Caswell, 1999). Until the invention of process-control methods in food production, biological contaminations were controlled by analysing the composition of a foodstuff, once produced or prepared, to discover the presence of any bacteria or contaminant. HACCP replaces this costly and post-incident control with a dynamic and preventive system that focuses the attention of in-house hygiene officers and external food inspectors on certain critical areas of the production line determined not only through testing but through more general knowledge of the typical contaminations and poisonings that occur in the food chain.

Despite a rather unpronounceable acronym (specialists say 'hazip'), HACCP has travelled widely – from an emergent industrial practice, aptly captured in very generic and portable principles, to regulation. The seven principles have remained remarkably intact as they travelled across time and space, to the point of appearing to be carved in stone. They emerged in an American company, and were then endorsed simultaneously by the FDA and the WHO. This transnational trajectory continued in the 1970s and 1980s, with HACCP being taken up by a variety of national food administrations as well as transnational professional and trade associations. Guidance was produced in various places as to how to apply these highly abstract and generic principles, capped by the enterprise to set up harmonised international guidance in Codex. The WHO and FAO placed the elaboration of a HACCP guideline on the agenda of Codex, the standards of which were soon to become mandatory in the framework of WTO agreements (Dawson, 1995).[2] Codex adopted a guideline for HACCP in 1993 and incorporated references to HACCP in food-specific hygiene codes from 1995 onwards. The European Commission proposed HACCP-based food-specific hygiene Directives in the early 1990s, to finally require food companies to adopt a HACCP-like system in the general food hygiene directive 93/43.

The adoption of a HACCP guideline by Codex is an interesting outcome in its own right because international standards have not always been seen as an appropriate instrument for regulating food. To begin with, it has always been difficult to define what a food is. Classifications have been established of course, as well as microbiological and chemical criteria of composition for each type of foodstuff. But standards that apply across all types of foods are rare. Until the Codex adopted the guideline, food hygiene was prescribed through food-specific codes of practice (except the so-called good hygiene practices). These were established for each type of foodstuff in close detail and included temperature and size of premises, tools and machines to be used, ingredients to avoid or timing of the introduction in the mixture, and so on. The whole trajectory of HACCP, the preservation of the integrity of the model over time and space, from Pillsbury to Codex and the Commission, is therefore quite extraordinary. It was very early on recognised as a valid and promising approach by the WHO and integrated into the 'basic texts' of international food hygiene rules (Motarjemi et al., 1996). It came to cover issues of biological contaminations as well as physical (harmful foreign objects, such as glass or metal fragments accidentally introduced in the food) and chemical risks (contamination by unwanted chemical substances). Finally, it also travelled geographically. It originated in the effort of large agro-food businesses from the west to standardise safety assurance, or at least to create internal benchmarks across all production sites. But it has been adopted by all countries internationally following its transformation into a legally binding international Codex and EU standard, and is now to be implemented by all types of food businesses including smaller hospitality businesses, primary producers and slaughterhouses.

Like PVP, the case of HACCP demonstrates the determining role that a scientific elite interested in safety and risk play in allowing international standards to crystallise. All along the history of HACCP, a set of international food microbiologists were involved and worked together to defend the principles and the integrity of the approach by establishing its positive contribution to the realisation of global regulatory and policy objectives. The particularity of HACCP is to result from initial modelling work carried out at and around Pillsbury, specifically by a food microbiologist called Howard Bauman, and continued through an invisible college that maintained the integrity of the model at the same time as they worked on applying it in various companies. They used this experience to confirm the benefits of applying the principles. They worked to secure these benefits by producing guidance on how to apply principles, and made sure their guidance over-rode the many others that appeared in the meantime. The adoption of the Codex standard thus reflects the continuous collective standardising work of food microbiologists who combined in a practical form industry experience, research capacity, close relations with international organisations and an ability to work together to promote a more analysis- and performance-based mode of food control, for their own benefits as well as for those of harmonisation-prone industry and standard-setting organisations.

THE HACCP PANACEA

Codex defines HACCP in the following manner:

> The HACCP system, which is science based and systematic, identifies specific hazards and measures for their control to ensure the safety of food. HACCP is a tool to assess hazards and establish control systems that focus on prevention rather than relying mainly on end-product testing. (Codex, 2003: 21)

This definition is now part of the 'Recommended International Code of Practice – General Principles of Food Hygiene' adopted by the Codex Alimentarius Commission. This document has legal status, and it lists the seven principles of HACCP, along with guidelines for their application. It reflects the thinking of scientists who promoted HACCP over the years and two arguments in particular. First, HACCP is not just random control. It focuses on what has been evaluated as being the most probable source of contamination, knowing the nature of the product, the types of ingredients, the different treatments they undergo along the production process, the conditions in which they get distributed and consumed, and so on. The HACCP acronym itself emphasises the embedding of control practices into a scientific attitude. It refers to the first two principles that embody this scientific component; that is, Analyse Hazards (HA) and establish Critical Control Points (CCP). It has also been argued that, ideally, data collected by a company as part of a HACCP system should be transmitted to governmental bodies and national laboratories to compute nationwide images of the prevalence and incidence of food contaminations. The data should then be retransmitted to food businesses to alert them to the most prevalent food contaminations on which their HACCP plans should focus (Notermans et al., 1996). The second element on which the preamble insists is that HACCP is a process-control method. The controls should be placed along the production line (Jongeneel and van Schothorst, 1994). There is no need to wait for the product to be finished to control it (e.g. check the presence of contaminants in it). This form of control is much more efficient and conscious of the constraints of production.

All of these positive arguments about HACCP are summarised in the preamble to the standard. According to this text, the HACCP model has the following properties:

> Any HACCP system is capable of accommodating change, such as advances in equipment design, processing procedures or technological developments. HACCP can be applied throughout the food chain from the primary producer to final consumer. As well as enhanced food safety, benefits include better use of resources and more timely response to problems. In addition, the application of HACCP systems can aid inspection by regulatory authorities and promote international trade by increasing confidence in food safety [. . .] The application of HACCP is compatible with the implementation of quality management systems, such as the ISO 9000 series, and is the system of choice in the management of food safety within such systems. (Codex, 1993)

HACCP by its model-like and generic nature lends itself well to the enunciation of the general properties presented here, which seem rather formidable. The Codex document is quite evangelical about HACCP, boasting about the policy benefits coming along with it, be it in terms of reducing the prevalence of food-borne diseases, minimising trade conflicts, avoiding or responding to food scares and indeed in terms of responding to all sorts of 'problems'. It reflects the immense opportunities in food regulation that such an instrument opens for the Codex – a body whose standards became legally binding with the adoption of the World Trade Organization (WTO) and Sanitary and Phytosanitary agreements (Garrett et al., 1998) – as well as for national governments and for the WHO, which repeatedly emphasises the above global policy challenges (WHO, 1997). One of the clear advantages of HACCP is to allow transnational regulatory bodies to establish standards on a matter that has long been difficult to regulate, namely food safety. The emergence of HACCP coincides with the development of international food trade as well as the awareness of the prevalence and incidence of food contaminations and food-borne diseases, particularly in the developing world (Käferstein et al., 1997). These issues have been on the agenda of the WHO and the European Commission. These institutions capitalised on the model-like nature of HACCP to build a generic response to this challenge, thus extending their intervention on food hygiene and food safety to a variety of settings, from primary production to distribution, storage or local preparation of food.

These properties derive from the set of seven principles, which have not changed since the early 1970s. The more specific content of the Codex standard concerns the guidance on how to apply these principles. The Codex document provides a set of self-explanatory but abstract actions to apply the principles and achieve expected benefits. It contains a diagram (Figure 5.1) that shows the logical order in which these actions should be carried out. In the standard, it appears as the 'logic sequence for the application of the HACCP'.

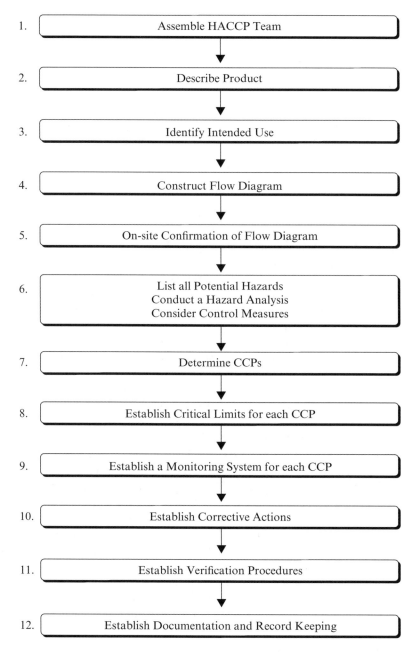

Figure 5.1   Logic sequence for application of HACCP (reproduced from Codex, 2003)

Each of these actions is explained in the standard, not in great detail however. Step 7, 'determine CCPs', is described in the following way:

> There may be more than one CCP at which control is applied to address the same hazard. The determination of a CCP in the HACCP system can be facilitated by the application of a decision tree, which indicates a logic reasoning approach. Application of a decision tree should be flexible, given whether the operation is for production, slaughter, processing, storage, distribution or other. It should be used for guidance when determining CCPs. This example of a decision tree may not be applicable to all situations. Other approaches may be used. Training in the application of the decision tree is recommended. (Codex, 2003: 26)

The explanations are thus not very operational, which confirms that the value of the Codex standard is more as a meta-standard, the essence of HACCP, to be able to evaluate other guidelines produced by government, professional or private bodies for HACCP.

The European Commission is another source of HACCP standards. The first European Directives related to food (Directives 64/432 and 64/433) concerned hygiene in production and preparation of meat, particularly in slaughterhouses and at subsequent stages. They were highly prescriptive and detailed. For various stages of the food chain, they summarised the points to which attention should be paid: such as the size of premises and types of construction materials, procedures for cleaning tools, room temperature, and so on. Following the same prescriptive logic, the European Commission produced more texts, each covering a type of food product. In the early 1990s, the European Commission prepared new texts concerning fish and fishery products. Directive 91/493 was the first to include a reference to the HACCP. The following year, HACCP was mentioned in Directives 92/5 on meat products and 92/46 on dairy products. Instead of inserting HACCP in the legislation product by product, the European Commission soon chose to adopt a horizontal directive for food hygiene. The HACCP is the instrument that allowed the creation of general 'blanket' prescriptions for the production of foodstuffs (Perissich, 1990). The generic and higher-order nature of the HACCP principles created a two-tier regime of harmonisation, whereby supranational bodies could set principles without having to prescribe technical details.

The European Commission, in the framework of the 'New Approach', welcomed a procedural tool like HACCP. According to this new policy adopted in the mid-1980s, the European Commission should not establish detailed technical prescriptions and specifications because its role is to articulate generic requirements for various classes of products relating mainly to safety and quality. The setting of particular rules for the production and distribution of each class of product or each specific product was delegated to standard-setting organisations. This new approach worked well for the European Commission in two ways (Egan, 1998). First, it dissociated political and technical components of regulation by preserving the policy-making role of the European Commission and protecting it from the failures of technologies. Second, it allowed standardisation and the free circulation of products across intra-European borders to progress in sectors that were thought to be too complex for products to be standardised. This was the case with food in Europe, where regulation of food had been hindered by the protection of national traditions of food production and consumption.[3] HACCP thus offered the possibility of regulating food more effectively than was the case under the old prescriptive approach. It enacted a sort of neo-liberal form of regulation by which the retreat of the main European executive body on policy-making roles and the enunciation of general requirements went in parallel with the progression of the scope of regulation (Borraz, 2007b).

The content of Directive 93/43 on food hygiene was discussed in the EU while the Codex was working on the HACCP principles and guidelines. The Directive is the first horizontal text on food hygiene and food safety and directly inspired by the new approach. It depicts food production as a chain of operations, at every stage of which, businesses involved should adopt HACCP plans. This prescription is valid for all products. The Directive makes reference to the Codex work on HACCP. In 2003, four new legal documents were adopted that further standardised and strengthened food hygiene regulations. Operators of all food chains are required to adopt the seven-principle HACCP plan in Directive 2003/53. The generalisation of the standard is correlated with the adoption in the framework of the EU General Food Law (Regulation 178/2002) so that any operator of a food chain is responsible for the safety of the products or substances it releases.

The formal and portable logic of HACCP also presented an opportunity to minimise the variations in modes of food control that originate in differing levels of resource and expertise across countries and businesses. A decision-aiding tool such as Figure 5.1 compensates for a lack of technical capacity or resources to operate control. It is premised on the belief that such formal logical tools are intuitive and universal. The use of these ideal-typical categories of situations and actions does not, in principle, require any specialist knowledge. In this sense, HACCP provides autonomy to businesses. Control becomes internal and self-control as much as control by official food inspectors. The company organises its own routine surveillance and correction of food contaminations. Governmental, intergovernmental and professional bodies provide more or less detailed guidelines on how to do so. In between industries and regulators, official veterinary inspectors are responsible for verifying the quality of HACCP plans set up by companies. Ideally, they act as consultants or advisers rather than as inspectors.

In other words, the HACCP model carries with it a strong potential for aligning all actors of food production and food hygiene, turning them into intervening actors of an overall food control system with global coherence and impact. The downside of founding food hygiene on such an abstract model is the difficulty in implementation however. There is particular difficulty for smaller businesses to align themselves, given that they do not support a high degree of organisational rationalisation and lack the capacity to put in place a safety process with dedicated resources and personnel. This was recognised incrementally in the Codex, where its third version of the HACCP guideline (adopted in 2003) admits for the first time that:

> Small and/or less developed businesses do not always have the resources and the necessary expertise on site for the development and implementation of an effective HACCP plan. In such situations, expert advice should be obtained from other sources, which may include: trade and industry associations, independent experts and regulatory authorities. HACCP literature and especially sector-specific HACCP guides can be valuable. HACCP guidance developed by experts relevant to the process or type of operation may provide a useful tool for businesses in designing and implementing the HACCP plan. (Codex, 2003: 24)

For all its simplicity the HACCP system entails a set of very demanding substantive prescriptions. In its logic, HACCP substantially changes the way food hygiene is managed by businesses. It replaces a system in which businesses comply with substantive criteria (including direction on the size of the building, height of the ceiling, cleaning instruments, contamination thresholds, etc.) with a self-regulation tool that requires expertise in calculating thresholds of contamination and probabilities of contamination, as well as the managerial skills to define and run quality assurance processes. It modifies the role of the inspectors from the control of compliance to providing advice and assistance for the setting up of HACCP plans (Pritchard and Walker, 1998). Ideally, HACCP inspectors now have the role of second-tier controllers, checking the content of HACCP plans as well as businesses' own biological and bacteriological analyses and corrections (Sitter and Van de Haar, 1998). Accreditation and certification bodies emerge as new actors in that configuration, with the obligation to audit and approve HACCP systems.

Such a formal process-based methodology of food hygiene is alien to the practice of smaller businesses in primary production as well as in hospitality and catering. A reorganisation of hygiene practices along the lines of a formal and logical system like HACCP is easier to achieve for larger companies with in-line production processes. These companies have staff who master formal systems, work with explicated internal production parameters (organisational structures, personnel competences, good hygiene practices) and implement strategic corporate objectives such as changing the relationship with food inspectors. These conditions are not always met. Operators who do not follow a linear production process find it more difficult to use the HACCP philosophy to organise their own control systems. Hospitality and retail businesses sometimes find it difficult to view their operations in the form of critical control points of an in-line safety system. They have little capacity to establish their own criteria for intervention, much less quantitative thresholds, often relying on food hygiene inspectors with regard to what to comply with and how. Instead of respecting technical and explicit numerical prescriptions coming from the 'top', they have to do their own hazard analysis and definition of critical control points. Furthermore, rationalising food hygiene under the form of a safety process appears of little added-value for businesses that are struggling in the first place to comply with basic hygiene requirements – the so-called 'good hygiene practices'.[4] The Codex guidelines are not easy to apply, and are particularly difficult for small users because they do not have a team of engineers and veterinarians to work with the abstract schemes. Many users appear to be at a loss when trying to cope with the different versions of the Codex guideline. Inspectors are similarly affected. Historically, in food law they have been given a sanctioning and coercive role but they have now been redefined by HACCP as auditors and consultants in internal safety processes.

Notwithstanding the simplicity and logic of its constitutive principles, in practice HACCP is a set of highly demanding prescriptions amounting to an obligation to invest resources in safety that are not easily met in practice. It has led specialists as well as governmental bodies to recognise that HACCP could never produce any of the hoped benefits unless practical guides were created (Mayes, 1992; Khandke and Mayes, 1998), training of inspectors and food businesses organised (Ehiri et al., 1995) and the actual effects of HACCP assessed (Unnevehr and Jensen, 1999; Motarjemi, 2000).

In reality, two phenomena could be observed along with the emergence of the HACCP model. The first is the proliferation of norms and support tools for food businesses. A dilemma emerged between the establishment of guidelines by supranational bodies for all operators of the food chain (leaving them the option of defining their own critical control points and contamination thresholds), or the establishment of specific food codes by professional or regulatory bodies, specifying the critical points and thresholds (Unnevehr and Jensen, 1999). This was the object of recurring discussions in Codex, where some members of the Committee for Food Hygiene pleaded for other Codex Committees (e.g. on Fresh Fruits and Vegetables, or Fish and Fishery Products) to carry over the HACCP principles into their own standards. National governments or professional associations intervene as intermediaries to help in enforcing the principles. The US Department of Agriculture is of the view that it must issue detailed regulations, laying down the list of critical control points, contamination thresholds and corrective measures for each type of foodstuff within official regulations. The European Commission decided to leave the details of the implementation of HACCP to professional bodies and to stay at the level of general principles. With increasing concentration, food retailers developed their own HACCP norms to impose on food producers (Dobson et al., 2003; Fulponi, 2006; Havinga, 2006).

Manuals for the implementation of HACCP principles have been created which explicate in full detail the meaning of each principle and the results that may be expected from their use. The International HACCP Alliance was created to help businesses in the meat and poultry sector to comply with the principles. The European Commission offered training programmes and technical assistance to operators in industrialised countries exporting food to the EU (Sperber, 1998a; Barnes and Mitchell, 2000). EU governments also amended the European Commission proposal to make HACCP mandatory for all operators in the food chain. They argued that primary producers would not be able to establish and run HACCP plans. The European Commission agreed, although the obligation remained for certain categories of smaller businesses such as small egg producers. Symbolically, the French Ministry of Agriculture subverted the issue by requiring small egg producers to establish some form of HACCP-inspired quality assurance system, but not to implement each and every principle. Egg producers are encouraged to identify the sources of contamination and the actions they may take if they detect a problem, but not to map out the production process and establish contamination thresholds or to keep records. In practice, food businesses, particularly the smaller ones, often need assistance from external consultants or Internet discussion groups. HACCP-based guidelines of food distributors, and other notes and technical studies produced by professional organisations on hazard analysis help food safety and quality officers to develop their plans. A large consulting and certification industry has also developed to advise food businesses on the setting up of HACCP systems. Paradoxically, the modelling of hygiene practices into an abstract and formal HACCP system indirectly legitimises the development of alternative norms and systems of food hygiene, which mimic rather than faithfully replicate the sequential application of the seven principles. Sub-regimes for food hygiene and references to different types of HACCP-like systems have proliferated to form a layered food regime (Caduff and Bernauer, 2006). The generic nature of the initial principles enabled various actors to prescribe their own food hygiene norms, as much as it fostered global harmonisation.

The second consequence is a decentralisation of the politics of definition of contamination criteria in businesses. While HACCP promised to promote rationally established CCPs that food businesses could copy from one context to another (Corlett and Stier, 1991), criteria in practice are negotiated locally between businesses and the inspectors. Identifying a particular stage of the production chain as a critical control point involves many disagreements between the producer and the inspector.[5] Faced with the complexity and comprehensiveness of the hazard analysis task (van Schothorst, 1990), food operators develop different sorts of strategies. Some put a critical control point in each room of the production site; others define as critical control points the items on which their clients demand satisfaction; others focus on well-known risks, such as the development of germs, and put a critical control point at every point where germs may develop. As a result, Bonnaud and Coppalle maintain, 'the establishment of HACCP, that is, of a methodology common to all businesses of all sectors, results in a multiplication of local scenes of technical negotiation between food industries and hygiene inspectors' (Bonnaud and Copalle, 2009: 415; my translation). The design of HACCP, while promising to re-centre food hygiene around businesses themselves and their capacity for self-control, also complicates the regulation of food hygiene, making its rules more numerous and more local. Returning to the origins of HACCP helps with understanding how food safety was modelled, in spite of the drawbacks of using generic and abstract forms of rules.

MODELLING HYGIENE PRACTICES

Reliable Food

The story of HACCP shows the power of concepts in transnational regulation: how one thing is turned into a regulatory standard as its constitutive properties are captured in a particular name. This is essentially what happened around HACCP.

The individual and professional ability of microbiologists to formulate models is fundamental to HACCP. As told in countless textbooks, academic papers or professional presentations, HACCP originates from the need to ensure that the foodstuffs used by astronauts during Apollo missions were 100% safe.[6] NASA was concerned that existing monitoring methods were imperfect. The safety of food was at the time verified through end-of-chain testing, a method that was inapplicable in space flights. With such a method, the probability that viruses, bacteria or toxins will contaminate the product can only be calculated after the event and is seldom reduced to zero. The foundational property articulated in this context is that end-of-chain testing equates with testing of the hazardousness of a product at one point in time, excluding the potential development of bacteria and other micro-organisms in the food at a later stage. There is, therefore, a need to establish the points in the production and preparation chain at which an agent that could cause contamination, even at a later stage, may appear. The concept on which HACCP is based thus does not define specific criteria of intervention. It only defines a highly logical protocol to use to define these criteria of intervention, named 'critical control points' in the official standard. This protocol allows definition of criteria at the local level, as close as possible to the point of preparation and use of the food. HACCP is therefore founded less on a central assessment of harms and risks than on the trust, typical of the 1960s, in formalised thought processes and systems thinking. This can be seen in the process that led to the invention of HACCP.

It was Paul Lachance, a biologist by training in charge of flight food and nutrition at NASA, who imagined that one solution to the limitation of end-of-chain testing could lie in the methods of reliability engineering to the production of embarked food products. Thinking in terms of fault tree analysis or other concepts such as 'critical failure areas' was widespread within NASA. It was applied to the design of the space shuttle, its components and also to weapons. The concept of critical control point applied in all programmes related to the Apollo mission (Lachance, 1997). Lachance requested the various contractors of the food programme, including the company Pillsbury, to measure and monitor pathogens in food. Applying the CCP concept was part of the food safety specifications.

Bauman, a food microbiologist by training, with experience in preparation of foods in submarines, led the team in Pillsbury that applied this specification. He enthusiastically pleaded for the application of this methodology in Pillsbury on all food production chains. At this stage, the seven principles did not yet exist. HACCP consisted of three elements: 'analyse hazards', 'identify critical control points', and 'monitoring'. Pillsbury is the place where two other principles were developed: establishing corrective action in cases of contamination, and defining limits of contamination. It was in 1971 at a conference of the National Association for Food

Protection that the first occasion to advertise the benefits associated with the concept of critical control points in the food industry arose, along with good manufacturing practices (Atkin et al., 1972).

Standardising the Concept

From this point onwards, Pillsbury and Bauman were seen as the proponents of HACCP. They were asked by the FDA to train its inspectors for the canning industry, which had problems with botulism in underprocessed canned foods in 1970 and 1971 (Bernard, 1998). As early as 1972, the WHO branded it as the best possible approach to food hygiene. HACCP was immediately recognised with a series of international and professional organisations advocating the adoption of its principles across the food industry. Indeed, different organisations converged towards HACCP and consistently reasserted the sequential application of these seven principles.

The history of HACCP shows an extraordinary continuity of the concept, with close coordination between supranational and professional regulatory bodies. Between Bauman's initial publication and Codex work, HACCP proponents worked to ensure that the concept would travel to all possible areas of application. The model was upheld and tested by Bauman and his colleagues. There is a clear trail of publications and working groups that repeatedly assert the integrity and coherence of the concept of critical control points, adding several principles to the initial three to form a complete system. This work also involved the development of tools and techniques that defined what a correct application and interpretation of these principles would be. Meetings were held in various places, the National Marine Fisheries Service, the National Food Processors Association, the FDA and the National Academy of Science as well as the ICMSF. Several of these meetings resulted in the publication of texts of guidance explaining the content, benefits and ways of applying HACCP to different food chains or products, all of them building on Bauman's earlier work.

The first of these organisations was the International Commission for the Microbiological Safety of Food (ICMSF), a small professional group of about 20 co-opted internationally renowned food microbiologists. Bauman was a member of this group. At the same time as the WHO food safety department declared food hygiene and microbiological issues more important than chemical ones, a first expert consultation at the WHO was held in 1972. Some ICMSF members there promoted HACCP as the tool for the WHO's policy to promote and build hygiene in food production. This led to a request from the WHO to ICMSF to clarify the principles and ways of applying HACCP. The result of that work was a seminal book published in 1988 (ICMSF, 1988) building on the first publications by Bauman (Bauman, 1974 and 1986). While only published in 1988, the book contained several chapters that had already been incorporated into the NRC's report of 1985 on microbiological criteria for food (NRC, 1985a). The second organisation to publish a text on the subject was the International Life Science Institute (ILSI), a foundation dedicated to food safety and nutritional issues and funded by a number of multinational food corporations. The ILSI utilised the expertise of those who had been involved with setting up HACCP plans in large multinational agro-food companies that were part of the ILSI. The work in the ILSI published in an ILSI booklet (ILSI Europe, 1993), drew on the experience of several food companies such as Nestlé. The ILSI developed a decision-tree (Figure 5.2), initially invented in Nestlé, which would eventually become a part of the Codex guideline.

Finally, the National Advisory Committee on Microbiological Criteria for Foods (NACMCF) also built on the HACCP concept to establish guidance for businesses. The NACMCF is a US interagency body that provides advice to food safety agencies and includes representatives of these agencies as well as of industry, academia and consumer groups. The NACMCF brought together microbiologists, food hygienists and food inspectors to produce its own recommendations in 1989. Its work represents an attempt by US regulators to spread their own version of a HACCP standard, and to create more uniformity internationally as HACCP started to spread around industrialised countries and was used without harmonised guidelines that companies could rely on (NACMCF, 1991). These texts were in their turn a synthesis of previous contributions. NACMCF recommendations used the experience of Pillsbury (NACMCF, 1991).

Each of these meetings and reports has also benefited from academic research and publications on the topic. Food microbiologists started to investigate the value of HACCP in various areas in the second half of the 1980s, the publication of the ICMSF book being a milestone in the constitution of this research field (even though the preparation of the book started much earlier in the year 1980). Academic research appeared in the journals of the field (*Food Control, International Journal of Food Microbiology*), starting from the early 1970s. However, the increasing number of publications did not make the field more varied. It stayed remarkably homogeneous, with hardly any author writing to establish the negative effects and failures of a HACCP approach. On the contrary, the initial building blocks established by Bauman remained and served to continue the elaboration of the HACCP system. The nature of the

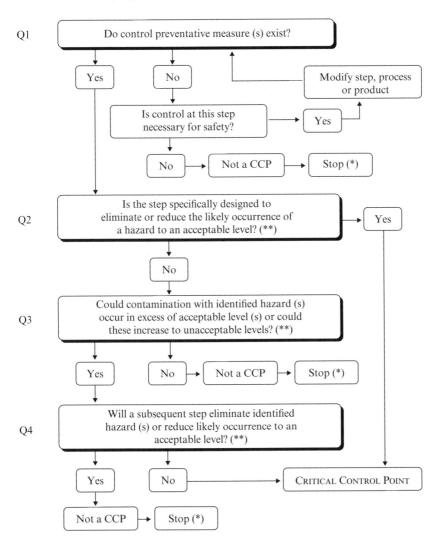

*Figure 5.2 Decision tree for the determination of critical control points
(reproduced from Codex, 2003)*

HACCP 'philosophy' was often reasserted (Panisello et al., 1999). The
adaptability of this 'logical approach' was also established by showing the
possible applications to a variety of products, physical and organisational
structures (Jouve et al., 1998; Holt and Henson, 2000).

Academic research also helped to maintain the model by demonstrating its logical and practical linkage with general management rules and

food law. Scientists for instance maintain that HACCP-based self-control systems are the best possible tool for producers to comply with the legal principle of primary responsibility for the products they release. HACCP records are sufficient evidence for businesses if they need to demonstrate before a court that they have taken all necessary actions to prevent contamination of foodstuffs, as required by the 'due diligence' rule (Blanchfield, 1992; Jongeneel and van Schothorst, 1992). Scientists defined how HACCP principles fitted with ISO 9000 (Harrigan, 1993), and later on with ISO 14000 systems. They demonstrated how HACCP matched the 'new approach' of the European Commission. They also contended that HACCP is an example of 'total quality management'. Most recently, they established similar linkages with risk analysis techniques for the calculation of critical control points. Finally, they showed how HACCP is a central element of what they call a 'food safety system', comprising all the institutions, policies, laws and guidelines that go into the regulation of food safety (Schilter et al., 2003). In their approach, data collected internally by companies should be centralised to allow governmental bodies to make better informed risk assessments, and food businesses should make use of more formal risk assessment schemes to define critical control points (Baird-Parker, 1995). These assessments in turn should help companies to focus on the most immediate dangers and prevalent risks leading to a global increase in food safety (Notermans et al., 1995). The scientific publications on HACCP have thus been remarkably consistent in maintaining and defending the model in its original definition, and upholding its properties.

Codex as a Receptacle of Expert Work

The topic of HACCP quickly appeared on the agenda of Codex, while other organisations such as the ILSI or ICMSF were working on the concept. It was first mentioned during the 1981 meeting of the Codex Committee on food hygiene. At this meeting, HACCP was presented as an approach (not 'the' approach) for dealing with the occurrence of pathogenic micro-organisms and their toxins in food. The report (Codex, 1981) states that the control over these micro-organisms must be exercised not only at the processing level but also during distribution, wholesale and retail storage and ultimate usage, either in food service establishments or the home. HACCP appears to be an approach that can be implemented at each of these stages. Codex however deferred to academic experts to develop guidance and specify the applicability of the concept for various food production chains, before going further in the development of a formal standard.

In 1985, consideration was given to the idea of revising specific product codes of hygiene to incorporate the HACCP concept in each. National delegations felt that the revision of individual codes of practice to include critical control points (above which monitoring must be done and corrective measures undertaken) would be an enormous task. Furthermore, local variations in the way foods are produced, 'even between two plants making the same end-product' (Codex, 1985), would be a barrier for applying HACCP across the board. The Codex codes of hygiene are also conceived of as general texts, which exclude such details as numerical critical control points. At this point of the discussion between national delegations, reference was made to the ongoing work of the ICMSF. The WHO indeed held regular expert meetings during these years, notably with ICMSF members who were working in parallel on what became a landmark book on HACCP (ICMSF, 1988) and key input for discussions within Codex. In 1989, it was logically decided by governments in Codex to only adopt the general HACCP principles and build on the work of 'competent authorities' like the ICMSF and the National Academy of Sciences. In the light of expert developments, it was decided that the establishment of 'general principles' was the correct way to proffer advice on the use of HACCP to businesses, as opposed to the update of product codes of hygiene with new substantive prescriptions. Obviously, this switch to a principle-based standard was an advantage for the conceptual work of scientists. It facilitated the further success of HACCP.

The Codex thus became the receptacle of texts developed by industry and academic specialists and a place to harmonise them, under the supervision of those who appeared to be expert in and keepers of the model. The UK and US governments volunteered to draft the 'working guidelines for the application of HACCP' and began to use experts' contributions by setting up working group meetings prior to the plenary meetings of the Codex Committee on Food Hygiene. The texts of the ICMSF, ILSI and NACMCF were jointly considered in Codex during an expert meeting at the training centre of the food industry in Chipping Campden, UK. The Codex guidelines clearly reflect the input of the organisations, for example through the inclusion of a decision tree for the determination of critical control points developed by Nestlé and ILSI. The national delegations within Codex worked on establishing commonly agreed definitions for each term of the HACCP model, and fine-tuning and naming the 'logic sequence' for the application of principles, as well as choosing a name for the overall standard (should it be called 'principles' or 'guidelines'?). These guidelines became an international standard in 1994 with the enforcement of the Sanitary and Phytosanitary agreement (the SPS agreement, which is the sanitary section of the World Trade Agreement),[7] shortly after the

European Commission integrated the standard in several directives.[8] The EU Directive 2003/53 eventually made it legally binding for all operators within the food chain to put in place a quality and safety assurance plan following the HACCP method, prompting the parallel development of public and private systems of support for the definition of critical control points and creation of HACCP plans.

In summary, HACCP as a regulatory model has made its way across boundaries – geographical ones, but also boundaries between types of food risks and expertise – to finally convince all actors that end-of-chain testing is, if not responsible for, at least too permissive of many possibilities of food contamination. But the translation of this model into a standard has been the subject of conflict between a formal and a conceptual approach, in which a generic protocol is supposed to facilitate the establishment of criteria of hygiene intervention. This is what food microbiologists and other proponents of HACCP have argued, and they said so because it has traditionally been part of their professional role to advise food manufacturers on where the chances of contamination are to be found. Other authorities have pushed for another mode of standardisation, in which it was still possible to set critical control points or criteria of intervention to help businesses. The intrusion of the HACCP model creates a dual situation. The generic model aligned major regulatory bodies and governments, as well as professional bodies and trade associations, enabling them to agree on a common set of typified practices and idealised policy achievements, by which food businesses, large ones in particular, took it upon themselves to contribute to the realisation of public goals and, in return, were relieved of the effort of complying with numerous substantive prescriptions. This was at the cost however of leaving discrepancies between the generic model and the local scripts developed to implement it, mainly because of the numerous intermediaries involved, including inspectors, food hygiene consultants and HACCP certifiers. The elitist nature of the group of food microbiologists who developed HACCP – elitist in their reorientation of food microbiology towards formal quality assurance and reliability thinking, in their collaboration with international organisations and also in their practice of consulting large businesses – does come as a likely explanation.

STANDARDISATION BY A DISCIPLINARY ELITE

It appears from such a narrative that there is no clear source of innovation for HACCP, except perhaps for the name. HACCP and its different principles crystallised through discussions on food hygiene rules and systems of control, which took place in a variety of venues across the world including NASA, Pillsbury, FDA, ICMSF and later on at the WHO, NACMCF, Codex and a few others. These sites were connected through scientists who circulated among them, and who worked consistently to give substance to the notion of critical control points by defining principles and elaborating tools for their application.

The Food Microbiology Elite

Participants in the HACCP odyssey noted retrospectively that 'An effective coordination effort [between] the regular food safety programs of WHO, FAO, the EU, ICMSF, ILSI, other groups, and the NACMCF [means that] national and international approaches to HACCP are decidedly similar' (Garrett et al., 1998: 179–180). This coordination was favoured by the fact that these organisations recruited the same elite food microbiologists to act as experts and give legitimacy to the guidance they created.

Bauman, for instance, decisively connected the first organisations in which a system for reliable food production was evoked (NASA, Pillsbury and the FDA) with the international organisation that would consistently promote it as a potential food standard: the WHO. Bauman was a member of ICMSF, the self-selected group of elite microbiologists who explicitly aim to influence the tools and norms of microbiological food safety. The diffusion of HACCP began with this group who relayed the work of Bauman in meetings with the WHO starting in 1972. The involvement of this group may be related to an enterprise of scientification and professionalisation of food hygiene. Food hygiene has for a long time been perceived to be a little considered technical job of applying unambiguous prescriptions of cleanliness and tidiness, sometimes involving tasks of sampling and analysis. Being technically easy to approach, food hygiene could also be carried out by other occupations than veterinarians and microbiologists. The willingness of ICMSF experts to depart from basic end-of-chain testing and to apply methods of reliability engineering specifically shows an attempt by food microbiologists to apply their capacity of inference – one of the three generic competences that define professional expertise according to Abbott, along with diagnosis and treatment (Abbott, 1988) – to food hygiene, transforming it on the way into a more general practice of quality assurance and food safety.

ICMSF members have this particularity to be among the main authors in the field and frequently take part in meetings organised by international bodies that the ICMSF, as a self-styled 'action-oriented' group, refers to as its audiences. From the year 1972 until the adoption of the HACCP guideline in Codex, the ICMSF functioned like a hub of HACCP experts available to work with the range of national and transnational regulatory organisations involved. ICMSF members have been seen collaborating with national governments (and their delegations to the Codex), but also with the NACMCF or ILSI. A younger colleague of Bauman, Michael, a Dutch microbiologist, also shows how central the ICMSF was as a college of HACCP experts. This Dutch microbiologist holds a PhD in veterinary medicine. Quite early in his career he became head of the national laboratory for zoonoses (diseases transmittable from animals to humans) at the national public health institute. He subsequently became head of the central quality assurance laboratory in Nestlé. He has been active within the ILSI, and participated in the expert consultations organised by the WHO and the FAO. He was a member of the Dutch and Swiss delegations to the Codex Alimentarius. Similarly, Jean-Claude, a French professor in food microbiology at the National Veterinary School who joined the ICMSF in recognition of his influential work on HACCP has advised the French consumer protection and agriculture ministries, and the European Commission. He also participated in ILSI meetings and has consulting activities. Institutions do not only use the ICMSF for its excellence. One of the experts enthusiastically explains that these organisations reconvened the group that was at work elsewhere in order to maintain the belief in the existence of a consensus:

> All the ground work is done by the WHO and the FAO. And they invite people with whom they have good contacts. That is why you find the same names! Because you want the best people. And they get good because they feed themselves. [. . .] United Kingdom, Denmark, Germany, Netherlands, and a bit France. Later on Belgium. Then the United States of course, Australia, New Zealand. And there is the whole mafia behind this, everybody talking to each other . . . always the same six heads. The advantage is that you can work very quickly.[9]

The accounts given by several protagonists estimate the pool of scientists used by the WHO, Codex and national governments at about 50 people, with a smaller circle of fewer than ten scientists considered to be the most authoritative on HACCP. Half of these 50 specialists and most of the shortlist of ten specialists have at some point in their career been part of ICMSF. Many of those who belonged to the 'mafia' that consistently advised governmental and intergovernmental organisations in the 1980s and 1990s are centrally positionned in the field of publications on HACCP. ICMSF members structure at least two central groups of co-authors and collectively stand as the most cited ones (see Figure 5.3).

The elitist nature of the field is illustrated by the slightly smaller number

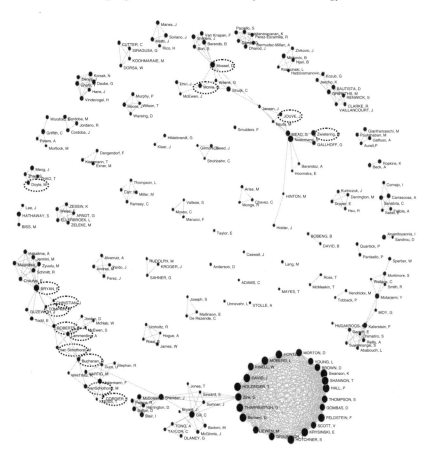

Figure 5.3   Map of co-authorship networks in the field of HACCP

of authors compared with the PVP map, and the presence of four dominant co-authorship groups. Strikingly, ICMSF members are dispersed in several of these groups that often correspond to sets of people working in the same institution, either a public health or national veterinary laboratory, a university department or a food company. Examples include small clusters on the periphery of Figure 5.3 such as the University of Liège in Belgium, the US Department of Agriculture, the University of Utrecht in the Netherlands, the UK-based scientific foundation Leatherhead Food Research, and the Microbiological and Food Safety Committee of the US National Food Processors Association (the tight 'ball' on the bottom right) – to create the longest chains of co-authorship displayed in the figure.

Of critical importance to the effectiveness of the invisible college is that all scientists recruited as HACCP experts by the WHO and national delegations to Codex also act as practitioners or professionals plying between sites of practice and sites of regulatory standard setting. ICMSF members advise or are directly employed by food companies. In that sense, they aim at the development of 'tools that help' (Mayes, 1998). Counter-balancing the abstractness of HACCP, they emphasise that they could only become HACCP specialists because they had worked on 'thousands' of cases of food contamination. In the same way, they would pledge that they would only develop HACCP plans for food production lines that they know. These scientists are among the few actors capable of relating the practical business of food production and hygiene control to the abstract and conceptual formulations of the HACCP standard. Conversely, the designing of specific tools for the application of HACCP (logic sequence, decision-tree for the determination of critical control points) owes a great deal to the practical experience of transnational HACCP specialists.

Recruiting Scientists, Manufacturing Consensus

A dominant logic of transnational regulation is consensual decision-making, and the recourse to experts who are cooperative with each other is instrumental in the construction of the consensus.

The authors who are known to be the best in the field are co-opted to become part of the ICMSF, and by way of consequence, to advise the institutions that seek support from the group. Using the ICMSF and, through it, the same set of experts, is a guarantee of credibility and legitimacy for organisations such as the WHO and Codex. As the above quote shows, it also increases the efficiency of meetings and discussions since it circumvents the lengthy discussions with Codex committees that are composed of national delegations. Forging consensus ahead of these meetings helps ensure national delegations converge more rapidly towards common positions. The manufacturing of a consensus involves more material acts of power, and the invisible college is set up because the institutions that it serves need to rely on scientists who consistently and coherently defend the same conceptual positions. On several occasions, some scientists who seemed to deviate from the definition of HACCP in seven principles, even with receivable arguments, were not re-invited to subsequent expert consultations.

The ambition to manufacture a consensus combines with the scientists' ideal of collegiality to explain why over the years the same body of scientists has been used by standard-setting and policy-making organisations.

For instance, the NACMCF was put together and given its mandate by the National Academy of Science. The ICMSF fulfils the function of a 'think-tank' (as described by Michael) for the WHO, which very frequently commissions the group or some of its members to produce reports prior to adopting its policies. National governments also recruit scientists and integrate them in their national delegations to the Codex. It is because of the close relationship and influence on WHO policies that members of the microbiology think-tank were recruited by their national governments, which guaranteed the latter a say in the regulatory system and control, where possible, over the drafting of international standards. It led to an overlap between the composition of the ICMSF, the working groups convened by the WHO and the FAO, and that of certain national delegations to Codex.

Their status as HACCP experts and contributors to the international success of this idea, in return, increased their feeling of solidarity. As hinted above, the ICMSF is a restricted group of about 20 scientists. One becomes a member by co-optation only. In the same way, being invited to meetings organised by international organisations is a key credential in this world of experts. It marks the difference between those who are part of the group of experts and those who are not. Over time, the recurrent collaboration of the same set of people under the aegis of international organisations creates an 'us-and-them' feeling. The group of scientists displays some cohesion and collaborative potential. These scientists share a common conception of their role as experts providing regulators the best possible tools for hygiene, and see themselves as constituting the professional elite who invented HACCP. The members of the ICMSF consciously form such a cadre. The frequent meetings with the WHO and Codex reinforce the cohesion of the group and the degree of interpersonal knowledge between them. Reinforcing the scientific and technical consensus that underpins the elaboration of HACCP is a shared working principle.

In return, the functioning of such a regulatory system that emerged because of the convergence of views on HACCP accentuates the faith and allegiance of experts and regulatory bodies in the concept. The publications of these scientists are remarkably evangelical about HACCP and the prescience of Bauman. Emphasising the life of the HACCP concept, the HACCP specialists see the end of a stage of development with the adoption of a guideline by the Codex and the start of a phase of implementation. Anything taking place now is part of implementation that cannot be blamed on the content of HACCP (Jouve, 1994a; Untermann, 1999; Motarjemi and Käferstein, 1999). Scientists thus rewrite the history of hygiene:

> It took nearly 50 years and the necessity of providing '100% safe' foods for the astronauts to get acceptance that line control in a systematic way is more reliable than end-product testing. It took another fifteen years for HACCP to get the recognition it merits. (van Schothorst and Jongeneel, 1992: 123)

Academic and expert work contributed to the creation of an ideal HACCP, and the concept seems to stand out on its own. In this configuration, and through a sort of teleology, experts tend to make it fully natural that the concept be recognised and standardised. Specialists write that the HACCP should be declared 'innocent' regarding its slow diffusion (Adams, 2002). The true reason is that it has neither been properly 'understood' (Adams, 2002) nor properly translated into different languages (Untermann, 1999). This tool is known to increase our control over food-borne safety hazards (Mayes, 1998) but has yet to make a full impact on the prevalence of food contaminations (Panisello et al., 1999) because it has been only partially implemented. This is a normal fact of life: any attempt to revolutionise practices is bound to encounter resistance (Jouve, 1994a). All of these factors should not overshadow the highly beneficial properties of HACCP systems, including its contribution to increased trust and mutual understanding (van Schothorst and Jongeneel, 1994). The fact that the prevalence of food-borne diseases remains high in spite of the adoption of HACCP in more and more countries and food chains is no proof of the ineffectiveness of the model. It is only that the model is recent and poorly implemented, especially in catering and hospitality businesses in developing countries (Motarjemi and Käferstein, 1999).

HACCP experts are thus united by a common commitment to push a model, the validity of which they believe in and that they further assert in their practice of on-site work. This relation to the concept betrays a desire to be effective in producing a standard as well as a form of allegiance to the expertise that is constitutive of their profession and research specialty. The preference for an abstract form of HACCP and Codex working principles is linked to interprofessional competition (Abbott, 1988). HACCP epitomises the intention of microbiologists to preserve their monopoly over food hygiene and to take up a position in the larger territory of 'food safety' for which they compete with other professionals such as toxicologists, nutritionists, and medical doctors. It also illustrates the attempt by veterinarians to cast themselves in a new role of 'food doctors' (Hubscher, 1999). Tellingly, the WHO announced that HACCP was the unique and best approach to food hygiene shortly after the take-over of the food safety department by a veterinarian in 1970. At the same time, hygiene and biological issues were declared to have priority in food safety over chemical contamination by the new head of the food safety department (a veterinarian who previously headed the sub-department of veterinary public health).

Working at the Larger Scale

It is striking that the circulation of members of the ICMSF contributed to the working together of a series of national and international standard-setting and industry bodies and to diffusing the HACCP concept within this circuit. This set of standard-setting bodies shared a generic model, but it was also disconnected, because of this very reliance on generic modes of standardisation, from local and sometimes national development of food hygiene rules and practices. The choices made in Codex, under the pressure of the WHO and allied experts, were reflective of this.

Codex had to follow the pace of experts' work on HACCP. Scientists have continually been active as experts at meetings of national delegations by recalling the state of knowledge on one or other item of discussion and replacing intergovernmental discussions within a process of constant refinement of a 'maturing concept' (Kvenberg, 1998). In doing so, they pre-empted the building of a consensus in the Codex arena that was the most receptive to scientists' constructions. Reports of Codex negotiations show that on several occasions intergovernmental discussions were on hold until experts made progress. On other occasions, experts also promoted this linkage between the intergovernmental work and their own agenda. Well into the 1990s and in Codex negotiations, the specialists argued that HACCP was still at an experimental stage, with more work needed for development (Motarjemi and Käferstein, 1999) and that, in spite of the adoption of Codex guidelines, 'further refinement' was needed (Mayes, 1998). Developing new research questions to spur the further development of the concept is of value in this.[10]

The European Commission was influenced in the same way. Bearing in mind the fate of smaller food businesses, the European Commission simplified the concept. Instead of seven principles, the 1993 General Hygiene Directive features only five.[11] It was only later, after many years of fundamental reform of European food safety legislation following the food safety scares of the 1990s, that the European Commission changed this. This first version was deemed too 'vague' (Untermann, 1999) and too 'implicit' in its reference to HACCP. The term-by-term comparison between the standard proposed by the European Commission and the set of seven principles of the 'official' HACCP (Jouve, 1994b) was influential in the European Commission eventually correcting its legislation. Its approach was soon reworked with the help of these very scientists who advised the WHO (Reichenbach, 1999).

Finally, the invisible college that turned out HACCP gave a larger place to the experience of multinational companies. Scientists' experience and expertise were linked to cooperation with larger businesses because they were more in favour of highly generic and 'liberal' forms of standardisation than smaller food operators, who required closer guidance on the hazards they faced and the critical control points they had to keep in check. The generic norm reflects the nature of the businesses or types of users with which they worked and the experience they drew on to develop the concept. This food microbiology elite acted as consultants for larger companies, or through the ILSI whose agenda is marked by the concerns of large multinational food businesses that fund the institute. This terrain of large in-line food processing companies took precedence over that of smaller businesses such as hospitality. Their emphasis on the generalisation of HACCP as a concept applicable across production lines and compatible with other general management standards and legal principles (ISO 9001, due diligence rule) also highlights the particular configuration whereby certain types of businesses were included rather than others. The local arrangements and, most importantly, the difficulty for smaller users taking up HACCP on the basis of Codex guidelines, reflect the nature of an invisible college and the degree of centralisation of a regulatory system, centred on big players that scientific experts worked with and implicitly favoured in their conceptual constructions.

The limits to HACCP are, conversely, linked to the exclusion of other potential users from this process. Limits on the capacity of smaller users to adapt HACCP thus result from the fact that they were less well represented in the invisible college and standard-setting arenas like the Codex Alimentarius. In concrete terms, HACCP guidelines remain difficult for them to handle and constant conceptual developments are seen as unhelpful. Thus some governments allow producers flexibility with regard to implementation.

CONCLUSION

The HACCP case illustrates the formidable capacity of mobile and polyvalent scientists to pre-empt the production of international standards. The highly abstract and portable model contributed to impose the perception that the overall prevalence of food contaminations could be reduced, following the experience and expertise of elite food microbiologists. The story of HACCP is unique in the initial and intensive work of modelling of new hygiene practices, as well as in the reiterated validation of this model by a college of scientists who were linked, corporatively and by their belief in the benefits of the model, to one of its principal initiators. The success of the action that started at Pillsbury is such that it is no longer possible to use a name other than HACCP to talk about food safety assurance. The concept has gained performative power. In itself it constitutes the reality of a systematic sequential food safety approach and is the accepted short name for it. Even food safety practices that do not result from the implementation of the concept can only be called 'HACCP-like' approaches. It was the circulation of food microbiologists, a majority of whom were co-opted and coordinated through the ICMSF, and of the model that allowed the WHO slowly to manoeuvre towards the adoption of a HACCP guideline inside the intergovernmental Codex arena. This circulation explains the close linkages between sites of evaluation of hygiene practices with standard-setting institutions. This circulation, much like in the case of PVP too, is reflected in the types of expertise used by HACCP proponents: legal and managerial knowledge, to judge from the roots of HACCP in decision-aiding approaches such as systems analysis, has here also been crucial to prove that HACCP has the formidable regulatory properties experts claimed for it.

The invisible college thus changed the regulation of food hygiene in two ways rather than one. It contributed to align its actors and formalise its practices. But it also, by way of consequence, rendered the conflicts and disagreements at the field level around what counts as a critical control point more frequent and visible. It made the interaction between businesses and inspectors complex. Experts like to underplay these conflicts and complications as inevitable. However, it also appears that they are the side-effects of standardising through generic and abstract means. The more generic and abstract the initial conceptualisation, the more issues appear in the implementation of the standard. Generic knowledge appeared to be the best way to persuade a set of regulatory bodies to create a guideline with a very large field of application. But abstract standards also call for intermediaries and support tools that undo scientists' constructions and displace the elite food microbiologists in their role of consulting and advising the food industry. Standardisation as a material reform of practices of control is limited in domains in which particular modes of control are less easy to establish as the typified cause of risks. Food microbiologists capitalised on managerial scripts to impose these principles and the drive towards more self-regulation and performance-based food hygiene, as much as on typifications of the lack of safety of foods. The HACCP case is thus different from PVP. Post-market monitoring of novel foods, very much an intermediary case between drug safety control and food hygiene, reveals other contrasts.

NOTES

1. Pillsbury is a Minneapolis-based food company, originally specialising in the production of flour and other baking products. The company expanded through merger and acquisitions after World War II into the processing and marketing of a larger range of packaged foods.
2. The Sanitary and Phytosanitary agreement provides that imported food products have equal treatment with domestic products. HACCP standards where they exist have to be followed unless the exporting country has scientific justifications for imposing different measures. With such products, the scientific case must be made following Codex standards for risk analysis. The choice of the Codex Alimentarius to be the source of the standards of references in the Sanitary and Phytosanitary Measures (SPS) agreement (see Büthe, 2009; Winickoff and Bushey, 2010 for analyses of this choice) had quite large implications on procedures of standard-setting, notably with the creation of science-based standard-setting scheme of risk analysis to preserve a mainly technocratic functioning while respecting public imperatives of transparency and democracy (Veggeland and Borgen, 2005; Post, 2006).
3. One of the most important European legal cases in the assertion of the free circulation of products in the (then called) European Community after all concerns an alcoholic drink (The 'Cassis de Dijon'case: Judgment of the Court of 20 February 1979 1 Rewe-Zentral AG v Bundesmonopolverwaltung für Branntwein (preliminary ruling requested by the Hessisches Finanzgericht) on 'Measures having an effect equivalent to quantitative restrictions', Case 120/78).
4. A HACCP plan in a company where good hygiene practices are effective would only comprise ten critical control points, whereas in another context up to a hundred critical control points may be required.
5. As noted by two researchers on the basis of a series of field-level observations, 'the discussion between the inspector and inspected does not come down to a dispassionate and conflict-free exchange of technical views. It is the occasion for each party to test the other's knowledge and expertise; it can include variable doses of bad faith, intimidation, irony, retention of information, etc. It is especially the case in slaughterhouses, in which inspection is the site of permanent power plays between inspectors and inspected' (Bonnaud and Copalle, 2009: 405; my translation).
6. It is likely that comparable food safety assurance systems were developed elsewhere (Mossel, 1989), notably in Japan. No sufficiently comprehensive empirical study of models and practices of quality and safety assurance in food have been conducted to prove that some HACCP-like system did not exist somewhere else. But, as this chapter shows, the standard adopted by Codex and the EU traces back to the USA. Its quite peculiar acronym must relate to the success of HACCP in terms of its capacity to travel. At the very least, it makes its genealogy (and the present work) possible.
7. For a state to block the import of a food product into its own territory, justification has to be made that HACCP guidelines were not adhered to or were insufficient to ensure food safety.
8. Directive 93/43 on food hygiene, Directive 91/493 on fishery products, Directives 92/5 and 92/46 on meat products and dairy products. At the time these were passed, most member states had required food operators to control food quality and safety, with examples being the UK's 1990 Food Safety Act and France's 26 September 1980 decree.
9. Interview with Michael, 10 December 2004 by phone.
10. In a speech concluding a food safety conference, a food safety expert with the US company Cargill and prolific author on HACCP thus argued: 'The aim of this conference was to improve the understanding of HACCP as a food safety management tool. One of the participants deplored that we raised more questions than answers. But I see that as a very positive sign. This conference shows how much effort is put into research – we have achieved progress while raising better questions to solve in the future' (Sperber, 1998b).

11. 'Analysing the potential food hazards in a food business operation, identifying the points in those operations where food hazards may occur, deciding which of the points identified are critical to food safety, identifying and implementing effective control and monitoring procedures at those critical points, and reviewing the analysis of food hazards, the critical control points and the control and monitoring procedures periodically and whenever the food business operations change' (article 3.2 of Directive 93/43). For the sake of simplifying the concept, the Directive breaks down the principle 2 of the pure HACCP into two steps, and merges the last four principles to form two simpler steps (control and monitor; review).

6. The value of abstraction: food safety scientists and the invention of post-market monitoring

In 2008, a group of academic and industry food scientists published a review paper in *Food and Chemical Toxicology*, an academic journal dedicated to toxicity testing and risk assessment of chemicals found in food, titled 'The application of post-market monitoring to novel foods' (Hepburn et al., 2008; the 'FCT paper' in the remainder of the chapter). The paper is a little more than an academic article. It was written by a group of scientists gathered together by the European branch of the International Life Science Institute (ILSI), an industry-funded scientific foundation. It gives the food industry long-awaited guidance on monitoring the risks of innovative food products in the market. Trivial as it may seem, agreeing on the elaboration, content and name of a guideline for the 'post-market monitoring' (PMM) of novel foods was far from a foregone conclusion. It took four years of intermittent work to write and publish this guidance, but it really took 25 years for its elaboration to be decided and its content to stabilise, following initial industry experiments on monitoring of the consumption and side-effects of novel foods.

There are cultural and political hurdles to creating such overarching regulatory instruments for foods. Novelty in human diet is not easy to measure. Diets are not systematically monitored, or not at the scale of ingredients and substances that compose foodstuffs. Any substance potentially has been consumed at some point in history or in one part of the world before its industrial use in food production began. It is simply impossible to know if that has been the case, for lack of information and data, in such a way that it is often difficult to reconstruct the 'history of significant use' that EU legislation commands, to know whether a food is really novel or not. Furthermore, it is hard to detect whether a substance has new or different effects on human health compared with another substance or ingredient. Existing methodologies are inadequate for studying the effects of foods on the human body. For instance, traditional toxicity tests are not adapted to whole foods. They are based on the principle of feeding laboratory animals with concentrated doses of a substance. And so far, the way to feed rats with concentrated doses of a whole food has not been found. The expertise in clinical trials for testing the positive health effects of so-called functional or health foods is also quite rare and few such trials are undertaken. There is no controlled distribution or pre-scription of these products, hence no systematic observation of the usage pattern and effects of these products.

The difficulty in regulating novel foods does not only lie in these techni-cal limitations. It also results from the political and cultural sensitiveness of our relationship with food. Can science mediate our relationship with food? Can food scientists such as toxicologists, biochemists, microbiol-ogists, exposure specialists, prescribe which products to consume on the grounds of 'safety' alone? Foods have traditionally been conceived as products that we are familiar with, that we intuitively know how to consume and are free to consume. The qualification of a substance or a complex of ingredients as 'food', and subsequently its ingestion, is some-thing that has historically been left unregulated. One way to distinguish a food supplement from a medicine is that the food supplement does not undergo such a strict and mandatory pre-marketing assessment as a medi-cine. Science-based regulation of foods, therefore, is a cultural battlefield. It is not evident that we need to test a substance to know whether or not it can be called a (novel) food.

In such a context, it seems all the more extraordinary that an agreement on protocols of risk assessment for novel foods can take form. And the reality is such that in this area standards have a limited scope of applica-tion. Underlying PMM is the idea that generalising consumer question-naires and setting up call lines for every product might not bring up anything more than 'nice to know' data, with no incidence on the occur-rence of side-effects of these products that are, in any case, marginal. And indeed the FCT paper sets restrictive criteria for the use of these protocols and methodologies. It also took time for the European Commission to insert in the food legislation a rule concerning the undertaking of post-market monitoring. In this light, PMM is a negative case of regulatory concept, one in which a college of mobile and poly-valent scientists made it possible to standardise a practice and produce a formal rule, but with minimal effects on the practice of regulating novel foods. It emerged from a college of scientists who could only mobilise limited practical experience of novel food safety – few cases, examined out of scientific and research interest rather than professional ethos, a relation to governmental and transnational organisations founded on scientific rather than regulatory advice and a collective work situated in an industrial foundation, with greater receptiveness to the goal of regulatory harmonisation than to researching risks. A college took shape thanks to the multiple positions held by a few scientists and their collective work. They also delimited it by agreeing to work within the limits of their experience as risk assessors of particular novel foods, so-called 'functional foods'. The restrictions imposed on the applicability of the standard reflect the composition of this college.

THE UNCERTAIN REGULATION OF NOVEL FOODS

Up to the early 2000s, the EU did not have a legal definition of a food product. The General Food Law (Regulation 178/2002) thereafter provided a common definition of food as 'any substance or product, whether processed, partially processed or unprocessed, intended to be, or reasonably expected to be ingested by humans' (art. 2). Integral to the definition of a food product is the fact that its consumption is *not* regulated, unlike other products with effects on human health, but partly remains or should remain a matter of consumer choice. One of the main criteria for defining food is that as a product or substance it caters for part of the physiological needs of a person, not that it has successfully gone through an evaluation of its safety, efficacy and quality as for medicines. Because of this cultural and political reluctance to intervene too closely in the qualities of food products, the creation of a category of novel food has been contentious. It impinges on the area of 'conventional' or 'habitual' food consumption, something that food law has notably been reluctant and unable to do (Tremollières, 2002). It also leads to the assimilation of a whole class of food products that is akin to medicines.

Novel foods can be characterised by several traits. They are developed by large manufacturers through knowledge-intensive processes of research and development. They appeared on supermarket shelves within the last two decades. The novelty often lies in their claiming a particular effect on human heath or including a substance not hitherto used as a food ingredient. And yet, these products are still foods. At least, their manufacturers would like them to be. They often resemble other well-known products, and are *used* like food, that is, they are ingested by the mouth during meals to feed rather than cure ourselves. That they are foods justifies not submitting them to safety tests before consumers can purchase them. The trials and tests they should undergo need not be as rigorous as those for medicines. Their composition, label, price, should be the target of regulatory controls, more than their effects, as is the tradition for foods.

However, the contemporary concern for safety and risks has challenged these classical modes of control. Being new, novel foods can be suspected to be unsafe. How innovative are these products really? How different are they from other products we know? What criteria should we apply to authorise them and what expertise should be used to decide on this? How and by whom should novel foods' safety be checked? Should they undergo mandatory pre-marketing tests, like pharmaceutical products, given their potentially dramatic effects? To what extent is it right to organise a surveillance of the consumption of these foods, in the same way that medicinal use is controlled by the doctor and the pharmacist? Is the knowledge gathered in experimental tests and risk assessments good enough expertise to base authorisation decisions on? These questions are repeatedly raised when food manufacturers seek to market such products as cholesterol-lowering yogurts, herbal tonic Noni juice, or pesticide-resistant plants. They are difficult to solve because they involve different cultural and normative conceptions of what constitutes food, in a context in which an approach to food as functional tends to standardise our relation with food and to make us think of particular diets as being local and inferior (Roberfroid, 1999).

Given the difficulty of defining generic expectations regarding what food should be to us, it has subsequently been difficult to impose the same evaluations and tests on all innovative food products or substances. It has been found much more convenient to compartmentalise food into myriad intermediary product categories. The complexity of the regulatory terminology has greatly increased in the last decades as new products have been developed to respond to new needs. Terms such as dietary supplements, functional foods, health foods, medical foods, foods for special dietary purposes, and so on were introduced, mainly out of industry's eagerness for the invention of new ways of marketing their foods (Nestlé, 2002). These terms qualify substances with very similar functions to cover the area of products that resemble food but are granted functions like medicines (see Figure 6.1).

'Functional foods' were distinguished from both foods and pharmaceuticals by scientists, but more rarely in law (Ashwell, 2003). In Japan, functional products were, for instance, required to consist of conventional ingredients or compositions and to be consumed in the same way as food, and as part of the staple diet.[1] Functional foods can neither be registered as medicines nor be allowed to be included in medical claims. They cannot address any disease or health risks outside reduction in blood cholesterol levels, the promotion of a healthy digestive system, increased resistance to disease, the promotion of healthy teeth and bones, and the provision of energy (Arai, 1996). These definitions simply endorsed the perceived needs of the general population. Functional foods became fashionable with the ageing Japanese society in the 1980s and the specific immune, endocrine, nerve, circulatory and digestive system problems of the elderly population.

The legal category to deal with functional foods in the EU remains the

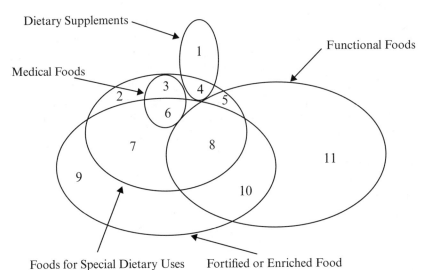

Figure 6.1   *Relationships between functional foods and other comparable*
                regulatory or commercial terms (reproduced from Kwak and
                Jukes, 2001: 110)

umbrella term of 'novel food' (Coppens et al., 2006). The construction of this category has not been easy. The difficulty of defining a category of novel foods and drawing a line between different sorts of products considerably slowed down the adoption of the 'novel food regulation'. Twenty years elapsed between the first attempt of the European Commission to put on the agenda the adoption of a piece of legislation for novel foods, and its adoption in 1997. This included a five-year period between the official proposal for a regulation and its final adoption by the Council – a rather long time given that the European Parliament was not involved in the adoption of this regulation.

The European Commission defended the view that novel foods can be generically defined when it proposed a European regulation for novel foods in 1992. These would be defined as foods not present in 'significant degree' in European diets or marketed in Europe prior to the enforcement of the regulation. And that, to the extent that our knowledge of these products cannot rely on a 'history of safe consumption', these products should be submitted to a mandatory pre-marketing evaluation. The legal regime is informed by a conception that these products are not well known, and are a priori unsafe.[2] EU member states preferred a partitioning strategy. They were of the view that novel foods should be defined as a collection of different types of technological innovations – genetically modified foods, plant-based foods, micro-organisms, and so on.

What was at stake in these different definitions was the more material reality of the tests that these products would undergo. A generic approach had the advantage, for the European Commission, of avoiding the lengthy discussions to define the tests and protocols to use for each type of product, the criteria of evaluation, the type of data to provide, at a time when the European Commission was delegating more technical tasks to standard-setting bodies and scientific committees. It guaranteed a simpler and more harmonised regime for businesses at a time when the European Commission showed concern for the detrimental effect on food innovation of national regimes developing in an uncoordinated manner.[3] The final text combines the two approaches, excluding certain products from its scope (food additives, and later genetically modified foods) but retaining the generic criteria crafted by the European Commission.

This generic approach created another sort of problem. It left this family of traditional products with an added health function without any clear definition, safety criteria or defined tests. Spontaneously, these foods have come to be defined by reference to what medicines can achieve – restore a health function or cure a disease – and to the way in which medicines are tested to prove these effects. It quickly became clear to the food industry and novel food scientists however that this comparison was not appropriate. Subjecting these substances – that were to remain foods and to be freely consumed products – to a safety assessment brought them closer to pharmaceuticals; that is, substances that are unanimously considered beneficial to health, but also inherently dangerous because of the adverse reactions they may provoke. The novel food regulation was thus born under the common expectation that tests and protocols of safety and risk assessment should not be as stringent as to incorporate novel foods with medicines, even in cases in which these novel foods carry health functions.

The FCT paper thus responds to the challenge of standardising a test for functional foods, without falling into an over-restrictive approach that would accredit the view that these foods require strict controls – such as specifying that clinical data should be provided, or that they should be distributed in pharmacies. It follows a fine line, creating a standard form of monitoring – hence one that can be extended to almost all functional foods, but also enriched foods and food supplements – but authoritatively reducing its scope. The authors emphasised that the test should not be generalised to become a systematic and mandatory control as pre-marketing safety tests are. Novel food scientists stood by the preferred approach of the European Commission to control foods through pre-marketing tests only. Scientists conformed to the view that foods cannot be overly controlled and PMM should remain an optional tool. Systematic surveillance post-market would go beyond accepted limits on the regulation of foods. They set three concrete limits to its use.

The first limit they set was that PMM should not be used to evaluate the efficacy of these foods that are often marketed with a health or functional claim. The decision to exclude 'functional efficacy' from the scope of the standard was debated within the ILSI group, but the need to fence off the domain of pharmaceutical regulation and differentiate the world of food from it appeared stronger.[4] PMM is not conceived as a tool to assess whether the functions advertised by the manufacturer are real. PMM is a safety tool, not one to measure the benefits of products. PMM should not be used to assess the effectiveness of novel foods, even though surveys and consumer call lines could very well be used to gather impressions of efficacy, as they are to collect warnings of side-effects.

Second, PMM should only be applied to assess those risks on which precise hypotheses can be formulated. However effective as a methodology to generate new data and fill knowledge gaps, risk assessors and decision-makers should discipline themselves to look at situations in which the 'need to know' is well specified. It cannot be used to collect information that is 'nice to know': 'PMM should only be used when triggered by or when the focus is on specific evidence-based questions' (Hepburn et al., 2008: 9). For instance, PMM may be justified in the following circumstances: when the ingredient is ingested in doses that are near the 'acceptable daily intake' (a threshold determined by toxicological testing and that is a hundred times lower than the dose at which no adverse effect can be observed); if one suspects that other groups than the population targeted by the manufacturer will purchase the product; or to provide reassurance regarding already existing knowledge about the frequency and gravity of a given health effect that the ingredient is suspected to cause. PMM may also be used to follow up on warnings registered through diet studies of adverse events caused by a given product.

Third, scientists defined PMM as a dispensable risk-assessment tool, rather than as the foundation of legally binding decisions. The ILSI guideline thus describes PMM as a complementary tool to be used as part of risk assessment, which is an advisory rather than prescriptive sequence in decision-making:

> PMM may serve as a complement to the pre-market safety assessment which is undertaken prior to the launch of specific novel foods and ingredients, in which case it is undertaken to meet defined, hypothesis-driven objectives and is solely related to safety aspects [. . .] PMM is NOT appropriate: - as a tool to replace any step in the pre-market safety assessment process. Safety should be assured before market-launch. (Hepburn et al., 2008: 28; capitalisation in original)

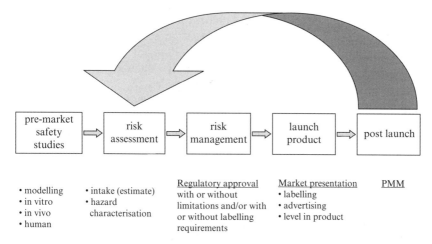

| pre-market safety studies | ⇒ | risk assessment | ⇒ | risk management | ⇒ | launch product | ⇒ | post launch |

| • modelling
• in vitro
• in vivo
• human | • intake (estimate)
• hazard characterisation | <u>Regulatory approval</u>
with or without
limitations and/or with
or without labelling
requirements | <u>Market presentation</u>
• labelling
• advertising
• level in product | PMM |

Figure 6.2   PMM: a complement to pre-market risk assessment (reproduced from Hepburn et al., 2008: 14)

The article also features Figure 6.2 with a very large arrow anchoring PMM in the pre-approval zone – although it takes place after the 'launch' of the product.

This approach contains a double restriction on the use of PMM. First, it means that it is for risk managers to decide to request a post-market monitoring study from a manufacturer, specifically because these studies concern products that have already been marketed or are about to be marketed. Requesting PMM takes place during the formation or revision of the regulatory decision, so it is only normal that those responsible for this decision decide on its follow-up. But it also means that the results of a PMM study will have to be assessed by scientific advisers, providing new information about the product rather than imposing per se a revision of the terms of the marketing authorisation. The FCT paper imposes two layers of precaution in the use of PMM, placing it in the ambit of risk management, post-decision, but anchoring it in the risk-assessment sequence. By this artificial depiction of PMM as a risk-assessment tool, the guideline emphasises that PMM is not a mandatory control on products. It should not be systematised because novel foods are not as likely to produce serious adverse events as medicines are. The scientists gathered in the ILSI also preserved the acceptability of the rule they were constructing and its potential to become a standard.

This article was part of a series of other pieces of guidance (Howlett et al., 2003; Schilter et al., 2003; König et al., 2004) on how to assess the safety of novel foods, all published in major regulatory science journals.

It provided a much-needed guideline for the industry and gave substance to a European Commission Recommendation according to which manufacturers should collect information on product consumption when their products can be expected to create substantial changes in people's diets and health.[5] Later on, the EFSA recommended the inclusion of PMM studies in applications for genetically modified foods, where appropriate (EFSA, 2004). The proposal of the European Commission for a new novel food regulation (Commission, 2008a) moved post-market monitoring up the normative pyramid, since it legalises the reference to this concept in Article 11.1:

> The Commission may impose for food safety reasons and following the opinion of the European Food Safety Authority, a requirement for post-market monitoring. The food business operators placing the food in the Community market shall be responsible for the implementation of the post-marketing requirements specified in the entry of the food concerned in the Community list of novel foods. (Article 11.1)

The Commission thus endorsed the terminology of post-market monitoring and implicitly designated the FTC paper – the source of this terminology – as the main guidance for it. The regulation regime for novel foods thus integrated the restrictive criteria of food safety intervention defined in this guidance. The concept underpinning the non-extension of PMM minimises the safety issue posed by functional foods. Going back to the elaboration of the guidance may help in elucidating this case of reluctant standardisation.

THEORISING FOOD MONITORING

Experiments

The domain of novel foods is extensive but knowledge of it is sparse. Experience in the regulation of novel foods is not equally distributed among member states. The Netherlands, the UK and Sweden have historically been preferred by manufacturers for the examination of their marketing authorisation application and to act as *rapporteur* for other member states on the European stage.[6] This has added to the early expertise of the agriculture or health ministries of these countries where the early businesses in developing functional foods were based. Novel foods are the result of a phase of research and development that has often taken several years; therefore there is a wealth of scientific data about these substances. Manufacturers frequently undertake, voluntarily, non-clinical and clinical tests as well as consumer surveys. However, there is no legal obligation for them to share this knowledge with public bodies. Against that scarcity, academic researchers are uniquely placed because they are exposed to industry tests and experiments. They are among the first to know or see the emergence of product innovations, carrying out tests in cooperation with these companies as consultants, buying small quantities of the new substance to carry out experiments themselves, knowing scientists employed by these corporations, having been in this position themselves previously in their career or evaluating those experiments and products as a scientific adviser to a public authority. It is via these channels that knowledge of products and safety issues potentially deserving of monitoring schemes has been accumulated.

The industrial development of novel foods started in the 1970s. Manufacturers immediately showed some awareness that they could have unknown adverse effects, and that these effects should be assessed. Conventional analysis of chemical structure and composition or animal feeding trials provided information in this direction. They were also conscious that the traditional controls for the safety of foodstuffs – labelling and inspection in sites of production and distribution – might be seen as inadequate by regulators because of their complexity, concentration and potential effects that made them more akin to food additives or even medicines, which undergo pre-marketing testing, than to 'conventional' foods.

The first form of pre-marketing control for novel foods was an agreement in 1980 between food manufacturers and the British Ministry for Agriculture, Fisheries and Food (MAFF). Despite the absence of an agreed definition of what a novel food is, the former would notify MAFF if they intended to put one on the market. One of the first novel products notified to MAFF in the early 1980s was Quorn, a product containing mycoprotein. Mycoprotein is a protein generated by a fungus that was isolated by the research centre of the food company RHM (Rank Hovis McDougall). The search for such a substance was initiated in the 1960s as a worldwide shortage of protein-rich foods and the spread of food insecurity mainly across Africa were becoming an acute problem particularly for the WHO and the FAO. The head of RHM, whose leadership was much inspired by his religious faith, considered it integral to his company's mission to contribute to solving this problem. R&D programmes were focused on the development of products that would provide the daily requirement of protein in just one portion.[7] One line of research pursued by the company was the development of foodstuffs from starch, in contrast to competitors that were then betting on single-cell biomass. This proved fruitful, leading in about ten years to the discovery of mycoprotein.

In 1983, RHM submitted a dossier of scientific data on mycoprotein to the Advisory Committee on Irradiation and Novel Foods of MAFF. At the time, there was no specific regime for novel or functional foods. Although the exchange of scientific information between the manufacturer and the scientific committee prefigures a procedure of marketing authorisation, it is voluntary and proceeds from a willingness to circulate information across a network of novel food scientists – academics, scientific advisers, corporate scientists – who are aware of emerging innovations and routinely communicate with each other. In this sense, tests conducted by manufacturers are never fully voluntary. They proceed from a 'need to know' that is perceived and shared by the manufacturers, the regulators and by scientific advisers. Often the tests on the product at that time were the object of informal negotiations between those three parties. In illustration, the development of Quorn was accompanied by a battery of tests on its functionality and safety. These included a form of monitoring of product consumers. In 1984, RHM made an agreement with several enterprise staff restaurants to distribute its mycoprotein-based product and, through questionnaires, to collect information from consumers. No concern arose from these tests and MAFF was duly notified about them.

As recounted later by a member of the scientific committee, the dossier did trigger a willingness to know more, as they judged the case 'interesting'. Furthermore, the members of the committee felt mycoprotein would not remain an isolated innovation. To them, it was part of a broader wave of innovation triggered by advances in genetic engineering although mycoprotein is not, strictly speaking, the result of genetic modification. The knowledge gained by studying mycoprotein combined with genetic engineering was used to differentiate types of innovation and data needs. The scientists of the committee put this into the form of a decision tree, each step in the decision being arbitrated by criteria that allow for distinguishing between product modifications, types of safety issues and information rubrics. Following the evaluation of mycoprotein and the advent of biotechnological food products, the scientific committee was renamed the Advisory Committee for Novel Food and Processes (Moseley, 1991). In 1988, having reached a satisfactory level of precision in the delimitation of novel foods from other foods, and established the ensuing methodologies and information rubrics to evaluate them, the UK made marketing authorisation mandatory for novel foods.

It was only much later that RHM's experimental form of field-level monitoring of food side-effects was formalised as post-market monitoring. This took place in the second half of the 1990s, following a new experiment by Unilever of 'post-launch monitoring' (PLM) of a cholesterol-lowering yellow fat spread (Lea and Hepburn, 2002). The Anglo-Dutch company began developing phytosterol esters in the mid-1990s. This substance had been found to reduce the level of cholesterol, which is a function that is normally ascribed to medicines. At the same time as Unilever was developing this food ingredient, pharmaceutical companies were investigating a new class of molecules, statins, which soon led to the marketing of blockbuster drugs such as Crestor or Lipobay (cerivastatin, see Chapter 4). In other words, phytosterol esters have effects that can be likened to pharmacological ones.[8] Anticipating this confusion, Unilever elaborated methodologies to establish the functions and safety of phytosterol esters. By that stage, Unilever had established a methodology for a form of monitoring of the large-scale consumption of the product and of emerging side-effects.

Unilever submitted the marketing authorisation dossier in 1998 to the Netherlands in the framework of the recently adopted EU novel food regulation. It was first examined there and then transmitted to all member states. Several national authorities raised the issue of the absence of controls over the distribution and consumption of a product where beneficial effects can only be secured if consumers ingest the right dose. Given these issues, the dossier was sent to the European Commission's Scientific Committee for Food (SCF) for a risk assessment of the product. In April 2000, the SCF confirmed that consumption of the product should in some way be placed under control. The effects of phytosterol esters do not increase beyond ingestion of doses higher than 3 mg/day, while the risks of reduced levels of carotenoid (pigments found in plants suspected of playing an important role in protection against chronic diseases) continue to increase proportionally. Furthermore, risks exist that consumers who neither seek nor need lowering of their cholesterol levels may purchase the product without recognising its particular properties.

The SCF judged that these issues did not justify *not* authorising the product and that the level of obligation placed on the manufacturer should not approximate that which exists for pharmaceuticals. Given these doubts, it was preferable for members of the SCF to grant Unilever a provisional marketing authorisation and request a post-launch study, the results of which would affect the renewal of the initial marketing authorisation. In its decision 2000/500 on the authorisation of the margarine, the European Commission requested that the study answers two questions: does the product actually reach the manufacturer's target consumer? Is the dose ingested daily by consumers lower than 3 mg/day?

The methods established by Unilever were then rapidly put into practice to glean answers to the questions posed above. The first method is that of the consumer panel. A small group of consumers are monitored, to see how many times they buy and use the product. The second methodology is the 'care line': a dedicated phone line with its number printed on the back of the product for consumers to report to Unilever. Employees in

the call centre are trained to question these consumers for usable information. This information is passed on to two experts of Unilever's internal
safety assurance centre: a toxicologist and a clinician. The chronology of
consumption and onset of effects is used as the prime marker of causality.
Where the two events are close to one another in time, one can suspect
causality. The two experts signal causality by a red light (orange being for
cases where the chronology indicates possible causation, but the effects
noticed do not belong to the known symptoms of the consumption of
phytosterol). When the two experts disagree, a larger committee examine
the case. This set of practices led one of the experts to state that they had
a proficient system in place and that they had become quite proficient at
it. The results of the study were seen as favourable in two respects. The
substantive results showed that consumed doses were lower than expected,
that the targeted group was actually reached and that there were no
adverse effects (no red signal was used: 148 calls from consumers out of
84,000 were marked with an orange signal). According to the EU's SCF,
this study was a reference in the domain and featured robust methodologies (Commission, 2002). Following this, the committee called in its report
for the creation of a guideline for post-launch surveillance.

Abstracting

The ILSI followed up on this request. The ILSI is an industry-funded
foundation, dedicated to the advancement of regulatory science, comprising scientific administrators of various task forces and working groups
made up of corporate and academic scientists. At such meetings, the
institute's policy is that it bears the participation costs of university-based
experts but companies pay for their own experts. The ILSI set up a novel
food task force in 1985 and a long-term member of the SCF headed this
task force at the time of the examination of the Unilever margarine. It was
under the aegis of this task force that an expert working group on PMM
was constituted, gathering together university-based and corporate scientists with experience in running PMM schemes.

What sort of exercise turned these experimental monitoring plans into a
standardised instrument with a public purpose? The working group started
its investigation in 2004. It reviewed PMM studies carried out across the
world since the 1980s, focusing on a handful of emblematic studies,
namely, aspartame (a sweetener), Olestra (a fat substitute used in savoury
snacks and developed by Procter & Gamble), a genetically modified form
of maize, and food products incorporating phytosterol esters.[9] All the
scientists involved had experience of monitoring either by examining companies' test results, or by conducting such tests themselves. The process

consisted of examining these four cases and the conditions in which PMM would be an appropriate tool. Within the ILSI working group, concrete, local, often mundane safety issues were conceptualised as information rubrics that PMM could fill. This was done by building on existing cases. Quite tellingly, the eight criteria of appropriate use of PMM formulated in the FCT paper are no more than the reformulation of the information needs that were inductively felt by regulators and manufacturers in the rare cases at hand. The first three criteria listed in the guidance paper are nothing but the three requests made by the European Commission to Unilever in 2000. For example, the guideline stipulates that:

> Where an original application was for one product, the exposure to a novel ingredient was limited. Further applications leading to several different products being marketed will have the potential to change the original exposure estimate. This may have consequences for the previously established acceptable margin of exposure which need to be monitored. (Hepburn et al., 2008: 13)

This criterion is quite simply the direct consequence of the Unilever case. Unilever initially proposed a margarine with phytosterol and later developed yoghurts and milk containing the same ingredient. It appears difficult, if not impossible, to stay below the safe level of ingestion of phystosterol esters that scientists determined, if someone consumes both a yoghurt and a portion of milk, for example. The multiplication of products containing phytosterol esters creates new health risks.

The scientists also turned the mundane matter of the control over the way consumers come to pick a particular margarine rather than another in a store, into a matter of 'consumer targeting'. The latter became a portable criterion to decide whether or not to do PMM. Where the consumer targeting potential of a product seems weak, such a study would be especially welcome. The same process was at work when scientists transformed the issue of the quantities spread by people on their bread into a notion of 'daily ingested doses'. In doing so, the scientists selected and changed a concrete matter into an abstract matter that could be controlled by one of the methodologies they developed. They created an assemblage between particular issues and an intervention, or problems and their solution. The temporal sequence by which their expertise and the guideline were created is telling: available methodologies, those tested by manufacturers such as RHM and Unilever, were considered first. It was their perceived robustness and results that convinced scientists and the ILSI that it was possible to craft a standard of intervention. In the ILSI then, the scientists described in abstract terms the safety issues that available methodologies could answer. In the final article, the trace of this intellectual process is erased. The recourse to PMM becomes the logical consequence of the need for information in the face of suspected risks. These safety issues have been objectified by scientists into action criteria. In the guidance paper, they became the 'triggers' for intervention.

In abstracting the limited situations in which practices similar to PMM were tested, scientists were guided by a shared understanding of safety arising from primary evaluations of risks and unofficial information. The perception that the development of a PMM concept was timely was linked to the knowledge shared by scientists of various specialities interested in novel foods, which was that the industry pipeline was about to generate a long series of health foods or functional foods. Heard more frequently in recent years was the scientists' view that biotechnological foods posed a lesser problem than suspected. Any adverse effect would have already emerged given the many millions of portions already ingested by animals and people around the world. The scientists seemed to share the presumption that biotechnological foods were not specifically problematic and that nothing should be done to amplify the problems that they are perceived to pose for the public (Hlywka et al., 2003). PMM therefore should not be made obligatory for genetically modified foods as it would otherwise lend credit to the view that this class of innovative product poses a specific safety problem (König et al., 2004). They also founded their work on the typification that foods, being more familiar and less concentrated than pharmaceuticals, are less likely in general to generate adverse effects and should not be as closely controlled as these products. This was reflected in the choice of the term PMM itself.

Scientists in the ILSI also considered the concrete issue of the technical investment required from manufacturers to do PMM, and the temptation of food regulators to emulate pharmacovigilance techniques and processes, when crafting the 'PMM' name itself. In practice, post-market monitoring is very similar to post-marketing surveillance (PMS). For instance, the techniques that employees are trained to use to question consumers closely resemble those that pharmacovigilance assessors use to obtain clinical information from physicians who report adverse effects. The use of chronology as the prime marker of causality is also directly imported from pharmaceutical safety. And in organisational terms, the division of tasks between routine evaluation by a smaller team of two experts and a larger committee for especially difficult risk assessments replicates the way drug safety regulatory agencies are organised, with junior people carrying out statistical analysis on all incoming reports, and teams of more experienced doctors looking qualitatively at more complex cases. To deflect accusations of mimicking pharmacovigilance, and of introducing in a less regulated domain the methods of a highly controlled one, Unilever experts decided to call the whole protocol 'post-launch monitoring' as opposed to 'post-marketing surveillance'. Later on, the ILSI working group chose to shift from PLM to 'post-market monitoring', with the same justification. In differentiating PMM from PMS, they aligned themselves with the interests of large food businesses that oppose the rise of regulatory requirements to the level of those in the pharmaceutical domain.

The risk-analysis language also served to draw a concept of monitoring from particular contexts, in particular regarding the process by which scientific advisers and bureaucrats agree to request studies from a manufacturer. The scientists gathered by the ILSI initially disagreed on whether to characterise PMM as risk assessment or risk management. When discussing with colleagues outside the ILSI, they encountered various views. Some argued that PMM was part of the management of risks: in temporal terms, PMM brings up new data after the launch and regulatory approval of the product. It leads to the revision of an already made decision and is an instrument for the bureaucrats who made the decision to authorise the product. Others would argue that it is part of risk assessment because it is not an obligatory measure, just an alternative protocol of safety information. But PMM offers an opportunity for risk assessors to review bureaucrats' decisions, as it is generally carried out after the decision to authorise a product. In the regulatory process, risk assessors intervene long after the initial risk assessment is terminated. By requesting PMM studies, risk assessors are involved, virtually at least, in the aftermath of the decision. In this sense, it is an instrument for monitoring both diets and marketing decisions. By monitoring the risks after the launch of products on the markets, risk assessors implicitly review the bureaucrats' decision to authorise a product. Reviewing is thus criticising, and involves a form of authority over those who made the decision. PMM hence creates a collision between risk assessors or scientific advisers and bureaucrats. And indeed scientists and bureaucrats have mutually accused each other of inappropriately requesting manufacturers to do PMM studies. Bureaucrats for instance argue that scientists are too often tempted to ask manufacturers for more data for the sake of research. They operate on a 'nice to know' rather than a 'need to know' approach. By abstracting these issues and reformulating them in the language of risk analysis, with its almost indistinguishable categories of risk assessment and risk management, the scientists found the beginning of a solution. Risk analysis does not solve the dilemma by prescribing particular practices (such as, e.g., classifying the effects into categories of plausibility and requiring bureaucrats to make their decision based on the number of effects of the highest plausibility). It leaves it to local users to find particular arrangements of this sort, encouraging more simply the interaction between the two.[10]

BOUNDARIES OF A COLLEGE, BOUNDARIES OF STANDARDISATION

What scientists in the ILSI really did was use a limited number of experimental monitoring schemes to validate and legitimise the commonly accepted view that monitoring should not be systematised when it comes to food. The PMM standard was constituted by the mere abstraction of situations in which it had effectively been used and was hardly extended as a result. Its small scope is a puzzling situation by which the capacity of scientists to produce very portable tools – what PMM actually is, since it could apply to enriched foods and food supplements indeed and in fact mimics pharmacovigilance – was curbed to adhere to a regulatory concept that emphasised the difference between functional foods and pharmaceuticals, the avoidance of too strict tests and the innocuity of novel foods. The reconstitution of the college in which this concept was articulated and transformed into a PMM standard highlights relevant explanations of this phenomenon.

Researchers, Scientific Advisers and Industry Advisors

The first condition in this process of legitimation by abstraction is, therefore, to centralise and cross-study what could be called a primary knowledge of risks and monitoring experiments. This is effectively what happened with the formation of a group of scientists within the ILSI as the institute only recruited scientists who had experience in the development and monitoring of novel foods. The capacity of the group to create a standard is linked to the centralisation of the primary knowledge of novel food monitoring within the ILSI.

The ILSI group comprised the few specialists who had been in a position to observe and experiment on novel foods and their safety at the field level. Those scientists, mostly toxicologists and biochemists, happened to be active scientific advisers to various regulatory organisations. Three individuals in particular may be singled out here as ones who established important connections between the work of the ILSI and novel food research and risk assessment. The idea of creating the working group was formulated by a British scientist, John, who was both a member of the EU's SCF and head of the ILSI novel food task force, which had already contributed guidelines on novel food safety assessment (Jonas et al., 1996). He was behind the proposal to set up a PMM working group within the ILSI after becoming acquainted in the SCF with Unilever's exemplary monitoring plan. He was also a member of the UK committee on irradiated foods when it examined mycoprotein in 1983. He is a regular delegate at the OECD and WHO meetings on matters of novel foods. When the EFSA was created in 2002, he became a member of one of its panels.

Two other French members of the working group have an interesting set of affiliations. The first one, Nicolas, is a biochemist at the French National Institute for Agronomic Research (INRA, or national centre for research on agriculture, environment and food matters). A specialist on the allergenic potential of genetically modified foods, he sits on the biotechnology expert committee of the French food safety agency. Through personal relations, he became involved with the work of Unilever on phytosterol esters and on the monitoring of the substances. He co-authored a paper with Unilever scientists on the utility of what was then called post-launch monitoring (Wal et al., 2003). He was a member of the SCF when it examined the Unilever dossier and requested the manufacturer to do a monitoring study. He then became a member of an EFSA panel when the latter was created. The second French scientist, Patrick, a medical doctor by training, is identified as one of the few specialists in the methodologies of measurement of food exposure to particular substances and nutrients. A researcher at the INRA, he is also simultaneously a member of the European SCF, of the joint WHO–FAO expert committee on food additives, of an expert panel of the French food safety agency and more recently, of a panel of the EFSA. Of course, the ILSI expert working group also mobilised scientists of concerned companies.

By their multiple positions and circulation, these scientists placed the ILSI within a circuit of novel food evaluation linking academic researchers, scientific advisers to governments and corporate scientists, drawing knowledge from this circuit to support a standard-setting effort. The ILSI played a crucial role in gathering together novel food scientists and articulating their primary knowledge to form a standard. The role of the ILSI was to delimit and provide a home to a set of scientists who interacted irregularly with each other outside the ILSI, mainly in scientific advisory committees. Figure 6.3 features a well delimited network of scientists who co-authored ILSI-sponsored papers on novel and functional foods, thus illustrating the contribution of the ILSI to the formalisation of a (previously invisible) college of scientists sharing an interest in the evaluation of novel foods.

The map does not feature many authors, which illustrates the fact that the exposure of consumers to novel whole foods and the safety of these products is an emerging field. There are only a handful of specialists, several of whom were part of the ILSI expert working group on PMM. The figure also shows a sort of 'fencing' effect: authors who are specialists in post-marketing surveillance of pharmaceuticals are not included in the cluster at the bottom which includes all ILSI-associated scientists.

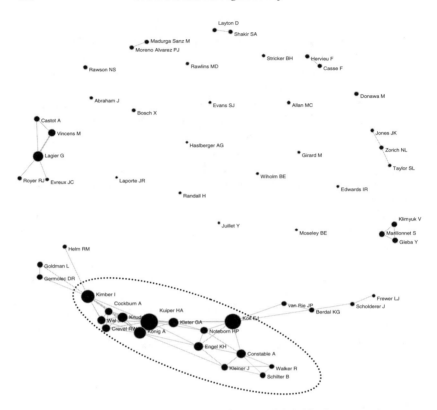

*Figure 6.3   Map of co-authorship networks in the field of post-market
                monitoring of novel foods*

However, the cluster includes specialists of both functional and genetically
modified foods. The network comprises ILSI scientists and EFSA experts.

   The working group was managed by an ILSI employee who had little
influence over the deliberations of scientists. Agro-food business actually
benefit from this hands-off position. It preserves the neutrality that is necessary to disseminate the guideline and present it as a standard, and allows
scientists to adhere to it and carry it around when they visit the many sites
of experiment and standard setting. This dissemination would be hindered
by a strict membership to the ILSI. Equally, the various mandates of
scientists as advisers to governments and others are not advertised. Their
participation in ILSI activities is consistent with their work as research scientists. This presumption of neutrality and objectivity is crucial. It allows
them to travel across sites and take on multiple roles across organisational
and professional boundaries.

Gathering these people together has helped the ILSI to take a central position as a standard-setting arena. Again, the circulation of John, Nicolas or Patrick is telling. They have participated in other meetings alongside or after their work with the ILSI, notably at the OECD. The OECD works by a similar method of collecting experience from member states and having delegates work together. It had begun collecting the experience and views of various member countries as early as 1993, as part of a reflection on the marketing of biotechnologies. A report stating that a consensus across member states was that monitoring in the market of biotechnologies could only be an instrument for testing well-defined hypotheses had been published (OECD, 2003) and was thus taken into account by the ILSI working group. The EFSA and the French and the British food safety agency all kept in touch with the work of the group thanks to Nicolas, Patrick and John who sit on their respective committees and were thus prepared to quickly take up the conclusions of the group. The accepted dependence of these bodies on the ILSI, through which the contributions of the industry to set standards is fostered, means that they integrate each other's work, rather than compete to impose their own normative production.

Frontiers of the College and of Standardisation

The place in which the scientists met and the many linkages they established around ILSI, with academia but also with the industry and international governmental bodies comes as a likely explanation for the preference to create an industry benchmark based on existing industry practices, as opposed to promoting PMM as an instrument to respond to a public problem of novel food safety.

First, these profiles mean that the college mediated a trio of actors: scientists (active as both researchers and risk assessors on the topic of novel foods), manufacturers (large ones in particular since novel foods are innovative foods generally proposed by multinational companies), public authorities (ministries of health, agriculture or consumer affairs of those countries concerned by the production and consumption of novel foods as well as the Directorate-General for health and consumer protection of the European Commission) and international standard-setting arenas such as the OECD. In other words, the college neither represents controversy-prone actors such as NGOs and consumer associations, nor such industries as food catering and food distribution, which may have been the bearers of alternative regulatory concepts for novel foods. On the contrary, its image of regulatory practice is built on the experience and agenda of producers of novel foods and most involved public regulators.

A second important factor in the restriction of the use of PMM is the limit of this college in terms of the experience and knowledge that were channelled within it. Standards are produced by those who have experience of the practices that will be standardised, and therefore depend on the way in which these actors interpret their own experience, technically as well as politically. The college, as gathered in the ILSI, is composed of scientists who engage with the regulation of novel foods as research scientists. Scientists involved in the evaluation of novel foods intervene as researchers rather than as professionals seeking to control the distribution and use of products. This is a paradox, given their many mandates as risk assessors. However, these mandates are coherent with their scientific careers as their action as expert or government adviser is part of a scientific specialisation that determines their career. The scientists depicted above have first entered national and then European, and sometimes international scientific committees. Specialism is the logic behind such opportunities. It is in their quality as specialist on an issue (the measure of exposure to food nutrients, food allergies, the toxicology of a particular class of food additives, etc.) that they hold such expert functions. If the work carried out as scientist and as expert is related, it is through content: following meetings of advisory committees, and taking in new research questions, hypotheses, references and sometimes data. They show some sensitivity to so-called 'applied research' or, in simpler words, they research questions posed by users of scientific information, rather than by academic peers. By focusing on a limited set of issues, scientists reassert the fact that they do not seek to fully control food production.

This limited involvement in regulation is also what their position of scientific adviser or risk assessor implies. PMM was defined as a risk-assessment tool because the technical expertise captured by the college was one of risk assessment. Scientists, as risk assessors, considered their legitimate professional role in regulation was in evaluating rather than requesting PMM studies. As revealed by their categorisation of PMM as a risk-assessment tool not to be manipulated by risk assessors, they show a sense of the limits of their legitimacy to define generally applicable criteria of safety of novel foods. They considered that generalising PMM would go too far, as it involves the assertion of generic assumptions about the risks of functional foods in general. They have therefore only gathered and reformulated the experience of food monitoring rather than pushed for its generalisation, staying within the limits of acceptable views of regulation of novel foods of the people and organisations they had most contact with; industries and national as well as European regulators.

The scientists belong, what is more, to food sciences. They share the conception that foodstuffs are not medicines – an expression often heard but seldom explained – and that novel foods despite their function should not be confused with medicines. By this conception they contest, rightly or wrongly, the perception that medical scientists are trying to appropriate the domain of food and nutrition. Food scientists thus defend the view that cholesterol-lowering has a 'physiological rather than pharmacological' effect. By claiming this distinction, they advocate the idea that food products could, in their own way, cure. They share the representation that there is a space for substances with possible health-enhancing effects that would be used by consumers outside any professional control.[11] They also express the accepted conception that foods should be less controlled than other products and that their role is not that of professionals aiming to control the prescription and distribution of these products. In short, foods are not medicines and thus food scientists are not medical doctors.

The last of the three contributing factors, which also derives from an understanding of the places and organisations among which scientists mainly moved, is that the production of the standard is a prolongation of a process of industry benchmarking, as opposed to a public response to salient adverse events or a research effort. The location of this standard-setting exercise within the ILSI, the aim of which was to influence the regulatory procedures for marketing food products, shows that the goal of regulatory harmonisation dominated that of the formulation of a political response to public controversies – not yet felt anyway in issues of functional foods. The idea of creating a guideline in the ILSI came from the observation by scientists of the SCF such as John, Nicolas and Patrick, that the industry needed guidance and that the work of Unilever was sufficiently convincing to be generalised. In turn, the preparation of a post-launch survey by Unilever was indeed aimed to set up an internal benchmark, in view of the fact that phytosterols would soon be introduced in more products and that, in general, the orientation towards more innovative health foods would become strong. The ILSI working group was in fact chaired by a toxicologist from the safety assurance centre of Unilever, who had also designed and run the post-launch survey. This also means that the overall ambition in the creation of such guidance was not to research risks, but to first and foremost establish a quality check on regulatory decisions, in the form of a harmonised standard that would contribute to clarify the regulation regime for novel foods, and ensure innovation continued in this domain.

CONCLUSION

Standardisation critically depends on the ability to form and delimit a college to standardise strictly the experience available within its boundaries. It leaves out of its scope the matters on which experience is not available, technically and politically. The strong emphasis on PMM as part of risk assessment shows how PMM standardisation conforms to an image of food regulation and of the limited legitimacy of scientists therein, in which the respective roles of scientists (who manage the boundary between research and expert functions in idiosyncratic ways) and that of bureaucrats can vary, and is often unclear. That the scientists adopted the terminology of risk assessment and risk management shows that they simply codified and reformulated this structure rather than aiming to change it. And although they showed a capacity to create a tool that could be applied in a multiplicity of cases, thus a tool that bureaucrats could take on and turn into an obligation for manufacturers, scientists respect a conventional distribution of roles that forbids them to go that far. By asserting that PMM is a risk-assessment tool (although used post-decision) to be applied in specific circumstances rather than a 'blanket' obligation, scientists established a restrictive criteria of PMM use to align all actors of the domain, in which they act as risk assessors. It was the limited experience of those who do research on novel foods and evaluate safety data – risk assessors – that informed the PMM standard. It was also the food scientists' recognition of their limited experience and role that made the standard an acceptable and successful one, effectively used as guidance by the industry in spite of its being just an academic paper.

The PMM case is a fascinating one in which a set of scientists managed to push forward a standard in spite of political uncertainties as to what the right tests, expertise and distribution of roles should be in controlling novel foods. This chapter looks into the college that allowed this to happen. Scientists, as they circulated between different testing sites or shared their experience, provided the primary knowledge to support the elaboration of a standard. In the context of food regulation, where science is one determinant of regulatory decisions among others, the articulation of such a concept is paradoxical. Scientists network regulatory actors into a system that turns out new rules but they cannot assume such enlarged regulatory power in their received role of 'risk assessor'. Although they designed a tool with many potential applications, they voluntarily restricted its use to limited circumstances in order to stay within the boundaries of their legitimate role in food regulation. They complied with the common representation that their experience is only one part of the experience on which regulatory decisions should be made. Their knowledge reflected the ambiguity inherent in the position of scientific advisers in matters of food safety. Although the college produced rules that could potentially enlarge the role of risk assessors to request data that is of direct relevance for risk-management decisions, they did not perceive it legitimate to take on such an extended function. Standardisation cannot simply be equated with the rise of new regulators. Scientists undoubtedly play a pervasive role in setting standards, but their own scope of intervention in regulatory processes did not expand as a result of the production of a standard that mirrors rather than changes the regulation of food safety.

The PMM case offers a contrasting view on the constitutive power of invisible colleges. The way PMM has been matched with general conceptions of what novel foods are is exemplary. It shows the particular ability of scientists to conceptualise experimental practices and to standardise them by coining abstract terms and elegant protocols. Furthermore, the way they linked together sites that are distant in geographical and regulatory terms is quite extraordinary. Finally, the convening action of an organisation such as the ILSI was instrumental in making these scientists work together. Scientists did influence the EU's novel food policy not simply because they put an already existing practice into a standard form but because they simultaneously tested this practice and represented the safety issues it could effectively solve. The constitutive power of science resides in the mobility and multi-professional capacity of scientists who are key agents of the elaboration of a common experience concerning monitoring and of its standardisation. The limited span of this circuit however also meant that scientists could only confirm existing usages rather than create new ones. Invisible colleges can only standardise what they network.

NOTES

1. Other criteria included the fact that there should be no significant loss in nutritive constituents of the food in comparison with those contained in similar types of food; that the food should be of a form normally consumed in daily dietary patterns, rather than consumed only occasionally; and that the product should be in the form of a normal food, not in another form, such as pills or capsules.
2. This directly results from the rise of food safety as a public problem on the European Commission agenda around 1996–1997. The BSE issue had turned into a political crisis that effectively contributed to the resignation of the European Commission headed by Jacques Santer and to the empowering of the European Parliament on these risk and consumer issues. Besides the BSE crisis, the Novel Food Regulation was adopted in the context of the controversy on the marketing of biotechnologies. The European Commission stood firmly on the adoption of a generic and formal criterion of definition of novel foods to install marketing authorisation as a regulatory instrument.
3. Novel foods first appeared in countries like the USA, Japan, Britain, and the Netherlands. Those countries started to develop their own legislation in the early 1980s.
4. This decision to exclude functionality from the scope was later confirmed by the adoption of a separate directive that organises the evaluation of functional claims made by manufacturers.
5. European Commission Recommendation 97/618/EC gives details of requirements of

scientific information and safety assessment report for application for the marketing of novel foods as established by the Novel Food Regulation 258/97.

6. In the European procedure for approval, companies send their application dossier to a single member state of their choice, which performs the first evaluation of this dossier and then, in a second step, acts as a 'rapporteur' towards other member states in the 'standing committee' that gathers them under the chairing of the European Commission.

7. This innovation policy towards food, which approaches nutrition through individual products rather than through diets, is the ideological component of the many attempts to give foods the form of pills, closer to what medicines are (products taken to achieve a determined effect in the body and with a given 'galenic' aspect). It also underscores the invention of the Mars bar (and its original marketing slogan: 'A Mars a day helps you work, rest and play') by Masterfoods, intended to cater in just one intake for all of a typical worker's daily energy needs.

8. All the more as phytosterols are present in the final product, such as the yellow fat spread pro.activ, at a level that is 8 to 15 times higher than in conventionally composed margarine.

9. Solid, unsaturated steroid alcohols (which occur naturally both free and as in this case in combination as esters) in plants. Phytosterol esters have been shown to reduce cholesterol.

10. The expressions of risk assessment and risk management are among those that serve to substantiate the limits of scientific evaluation in contrast to decision making (see Chapter 1). They help scientists to articulate the differences between a scientific activity and a political one. This recognition that science was politicised and politics scientific in an inseparable way in the original NRC report and the continued use of these categories, show that in the food safety domain, scientific experts did not necessarily believe in their right and legitimacy to make decisions. Interestingly, one of the recommendations of the report i.e. the creation of guidelines, has not been followed up in the USA or in Europe (Rodricks, 2003). This could have been an opportunity for scientific experts to design institutions that would be most favourable to their involvement. Risk analysis functions instead as a language used by both scientists and bureaucrats to justify constant reclarification of the boundary in a decision-making process. It also serves outsiders like activists who denounce and circumscribe the role of experts which they see as too industry-friendly in a smaller perimeter. It also works as a guideline on how to orient institutional reforms such as those following revelations that scientists took too large a regulatory role in the management of mad cow disease (BSE).

11. This was contested by the Swedish food administration whose scientific experts claimed that the Unilever product should have been examined as a medicine, and by a body competent in assessing pharmaceuticals rather than foodstuffs.

7. Exploring invisible colleges: sociology of the standardising scientist

Regulatory concepts emerge thanks to the circulation of scientists, the evaluative work they carry out in a variety of sites individually or collectively, and the circuit of standardisation that they shape linking together multinational companies, professional associations and learned societies, government, and intergovernmental and supranational organisations. Pharmacovigilance planning moved closer to being an international standard thanks to two successive moves, from the MCA working group (which channelled a number of similar concepts into a clear formulation) to the CIOMS and then to the ICH. The HACCP story is to a large extent that of a relatively stable set of scientists, most of them active through the ICMSF, WHO expert consultations and national delegations to the Codex Alimentarius. The emergence of a unique post-market monitoring standard was helped by the gathering in the ILSI of a small group of industry and academic scientists who knew each other through their activities on scientific advice, respective publications or previous ILSI projects. These stories thus illustrate how intimate science and transnational risk regulation have become. They show how scientists, standing as experts in multiple sites of food or medicines control, forge concepts that align actors and render standard-setting simply possible. The discussion of the concepts of communities and networks in Chapter 2 shows that the influence of transnational scientific experts on regulatory policies is not a newly recognised phenomenon. Scientists' specific ability to develop ideas, their habit of collegial deliberation, the elements of knowledge they share when their research deals with similar topics or when they are from the same discipline, all constitute a form of technical diplomacy (Hawkins, 1995) that is highly productive in setting standards. And yet, there are many different ways of defining an expert depending on the context or the view that is adopted (Nunn, 2008).

This is, first, because regulation is a complex system, with a variety of actors and processes involved. When scientists are involved in regulation, at what level really do they get involved? In whose institutional enterprises are they participating? With what institutional identity? Through which formal organisations? In the above cases, scientists are participating in regulation through not one but multiple vectors, ranging from scientific advice to private consultation and membership of governmental delegations in international organisations. If all the scientists presented in previous chapters took part in a common enterprise of standard setting, this does not mean they did not have other activities and roles in regulation at other points in time and on other issues. Second, science means a whole lot of different things too. Scientists may have different conceptions of what differentiates research science from evaluative science. To some, there is no difference as their whole discipline is evaluative in aim, and serves to regulate. In this case, the entirety of their research is relevant to regulation. To others, science should stop where decisions are being made about exposing consumers to a product. If they take part in regulation, it is because they are specialists on a product, substance or particular adverse effect. Their activities as experts are linked, if indirectly, to experimental research. Scientists who take part in regulatory operations may have a different conception of the civic dimension of science. To some, involvement in regulatory issues promotes their discipline and to others, it is a contribution to solving public problems. Even those scientists who work in the same committee or regulatory agency may have distinct ways of approaching regulation. Finally, there may be a substantial difference between a career-long involvement in a series of scientific committees on the one hand and one-off participation in such a committee. In one case, expertise could really denote an occupation parallel to research activities and as time-consuming as the latter. In other cases, it is close to being just a rhetoric that scientists play with to legitimise multiple commitments. The third element that raises questions about the identity of scientific experts is the finding that emerges from the three previous case-studies. This is the fact that conceptual work takes very different forms, and is less beneficial directly to the experts who were their proponents in their professional role of assessor than we might have thought from following, for instance, the notion of sociologists, theoretically applicable to this context of regulation, that the codification of knowledge and methodologies into abstract forms is a way to extend a jurisdiction, by making competences and tasks less accessible to competitors (Abbott, 1988; Freidson, 2001). Previous chapters showed that the logic behind the articulation and standardisation of regulatory concepts was less one of competition for more space and power in regulation, and more a form of intermediation and integration of regulatory actors through common knowledge.

So what does characterise the scientists who were active in developing and promoting PVP, PMM or HACCP? What do they have in common that explains their coming together to theorise and standardise regulatory practice? This chapter revives the notion of invisible colleges and applies it to the activity of these experts in order to come up with novel descriptive elements about transnational scientific experts. This chapter uses data on the protagonists of the preceding stories including biographical data as well as information about the formal and informal ties between experts, to inform this description of invisible colleges. It shows that these scientists have in common the ability to be an elite in their field of publication; that they have always worked across professional boundaries, in a way that is self-reinforcing as their career became transnational. Finally, and even though these scientists are linked together by weak ties, the strategic orientation of their interactions, notably the face-to-face interactions held through interstitial organisations like think-tanks, foundations or consensus conferences, leading to the often successful setting of international standards, results in the recognition of collectives (the invisible colleges) and for the individual scientists the capacity to assume the regulatory dimension of their expert identity.

CIRCULATION AND MULTI-PROFESSIONALISM

The case-studies all illustrated the fact that the invisible colleges through which the concepts were standardised comprised small numbers of generally widely published scientists. Pharmacovigilance, novel food safety and food hygiene are research areas in their own right. Potentially large populations of researchers are interested in these topics. Narrowing the focus to the set of researchers interested in pharmacovigilance planning, post-market monitoring and HACCP, one still discovers rather large populations. According to the results of a search in the ISI web of science, 224 researchers had authored more than one paper on pharmacovigilance planning and related themes in 2010. The numbers are 186 for HACCP and 450 for the monitoring of novel foods.

The overall finding that emerges from the three maps presented in previous chapters (Figures 4.2, 5.3 and 6.3), concerning all three cases, is that scientists who were part of invisible colleges are among the dominant authors in their fields. The maps confirm that scientific experts constitute the elite of evaluative sciences, in the sense that they are both among the authors who publish the most in the field and have the capacity to mobilise other scientists who are in the same position to either co-author papers or work together in expert meetings. They also confirm the distinctiveness of each case. The fields corresponding to standards with higher impact (PVP and HACCP) are more dense and competitive. The ability to create

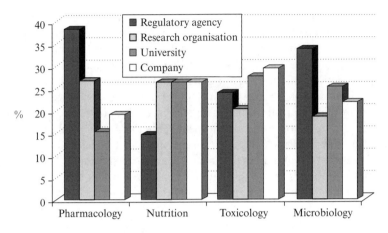

Figure 7.1   Employers of scientists authoring papers on PVP, HACCP and PMM per discipline

structure is critical in standardisation, and is the mark of a high level of centrality, in academic terms (number of publications, number of citations) but also in regulatory terms when these fields are professionalised and the scientists also occupy roles in regulatory intervention (in decreasing order: the pharmaceutical domain mainly, with doctors contributing to control medicines via prescription and notification of side-effects, being also among those qualified to become a 'pharmacovigilance responsible person' for the industry; with the food domain coming next, where food microbiologists and toxicologists act as risk assessors or as food safety consultants with the industry).

Is this elite characterised by a specific position in regulation? The first thing to note is that the participation of these 'evaluative scientists' in regulation in general greatly varies within and across domains. Evaluating risks does not signify working for any specific type of organisation, be it government, industry body or regulatory agency. Figure 7.1 exposes the affiliations of the authors presented in Figures 4.2, 5.3 and 6.3. Each scientific discipline has its specific profile, but overall, it is by no means possible to equate a status as expert in evaluation to a particular organisational affiliation.

The four affiliations calculated in Figure 7.1 by percentage are (from left to right, in each discipline): experts employed and paid by a regulatory agency, by a research organisation, by a university or by a company. Experts in pharmacology are more often based in regulatory agencies than other types of organisations. To a lesser extent, this also applies to experts of microbiological food safety. Food safety, animal health agencies and associated official laboratories offer a large number of employment opportunities. Nutritional assessment however has yet to attract intensive regulatory activity. Apart from the assessment of nutritional and health claims, nutritionists have little specific work to do in government agencies as they offer few employment opportunities. But nutritionists can be found in greater numbers in institutes, including private ones that develop and test new food products. As far as toxicity testing is concerned, experts are distributed rather equally among organisations. Toxicologists are mainly in chemical companies or research institutes, with a number of them being based in intermediary organisations such as contract research organisations which perform tests and other investigations for a regulatory purpose on behalf of chemical manufacturers. Similarly, microbiologists like toxicologists are not based in universities in great numbers.

These data illustrate that scientists participating in risk regulation are more frequently based in an organisation in which activities of evaluation create jobs and generate staff. And these organisations are not necessarily of the same kind in each domain. The dominant type of organisation is that in which relevant disciplines have been historically based; that is, within or outside universities. This is now being modified as independent regulatory agencies have been created in the domains of drug safety and food safety covering microbiology, toxicology and to a lesser extent nutrition. But this trend does not affect the affiliations of scientists to the point of creating a convergence across domains. Experts are not scientists who are based in a particular type of organisation. Regulatory science, from this point of view, is undoubtedly a hybrid (Irwin et al., 1997).

What then would characterise those scientists who collaborated together to theorise and standardise regulatory practice, as the different stories here illustrate? Rather than affiliation to one particular type of organisation, what defines actors of invisible colleges is their circulation and their accumulation of positions and projects simultaneously or successively during their career. Jasanoff thus notes that 'expert' is a label that allows scientists to act in multiple capacities as technical consultants, trainers, peer reviewers, advisers, judges, and so on (Jasanoff, 1990a) – 'hyper-experts' indeed (Mercer, 2004). These experts can specifically be defined by their acceptance of a variety of commissions and activities that lead them to act as scientific advisers, consultants to industry, national delegates in intergovernmental bodies or advisers to supranational institutions, beyond their role as academic or researcher. The ability to accumulate missions and commissions with a variety of organisations across the regulatory domain seems to be the preserve of a smaller set of scientists within that broader population, the question being what sparks this circulation and gives the necessary resources for it?

The occupational diversity just observed in the larger population of those who write about regulatory concepts is replicated at the level of the smaller population of invisible college scientists. These scientists are not affiliated to the same organisations. There is no possibility of explaining their engagement in regulatory conceptualisation and standard setting by the fact that they belong to a particular organisation. First of all, there are differences between domains regarding the jobs of these scientists. In the case of pharmacovigilance planning, two are assessors in a regulatory agency, the third is a 'pharmacovigilance responsible person' with a large pharmaceutical company. In the case of novel foods, all three experts do research or teaching for a living. In the case of HACCP, one expert is based in academia, the other in industry. Second, it is impossible to put forward one type of affiliation as giving access to regulatory influence because each of these experts has changed jobs in the course of their careers, sometimes in the middle of their work on the standards that are attributed to them. Two medical experts have moved from being hospital physicians to a position with a regulatory agency or with the industry. In the case of HACCP, Michael moved from being a researcher to a position in the food industry in quality assurance functions.

What however seems more consistent across the different personal situations is the accumulation of memberships of different advisory committees, in continuity with and thanks to their main occupation, which is the starting point for circulation. Paul and Waller extended their jobs as assessors at the MCA by memberships in European-level expert committees of the EMA. The specialists of novel foods have during their careers spent more time in scientific advisory committees examining industry applications and assessing risks. This combination allows them to maintain links or collaborations, not very visible ones though, with the industry. Being an industry safety officer does not preclude this sort of dual role either, since Gabriel participates in relevant working groups of the trade association of the industry he works for and even, as specialist of pharmacovigilance, with the regulatory agency of his country. In food, all scientists begin participating in activities of one committee on the basis of their speciality. They then progressively move away from this vision of expertise as a prolongation of their research specialty as they join other committees and gradually acquire particular skills as scientific advisers. The latter include synthesising research findings, bringing on board views that differ from one's own interests and favourite approaches, translation of quantitative and technical indicators into qualitative statements, collective deliberations, writing of opinions, and understanding the political and legal dimensions of technical issues. Scientists learn how to formulate policy options that are acceptable for decision-makers and that, at the same time, preserve their credibility as scientists (van Eijdhoven and Groenewegen, 1991). These combined skills turn scientists into regulatory consultants able to advise a range of policy-makers beyond the confines of their research. Strong continuity in the participation of committees can indeed be observed for most scientists as well as concurrently holding several terms of office, for example in a national and a European or international committee.

What is distinctive then is the combination of belonging to a scientific elite as seen above on the maps, and being in the position of an intervening actor in regulation (as an employee of a regulatory agency, scientific adviser to government for licensing decisions, an industry safety officer, etc.), the combination of the two resulting in sustained participation in governmental and industry advisory committees or working groups, as well as in this evaluative perspective that defines them intellectually.

This combination of positions is more or less easy to render coherent depending on the level of professionalisation of the discipline they are in. By professionalisation is meant the formal or informal recognition of the necessity to possess a particular knowledge to access a position. Professionalisation is strong in pharmaceutical evaluation. In the area of medicines, most scientists involved in the production of regulatory standards belong, by training and their curriculum, to the medical profession. Within that, the experts who participated in PVP and other comparable standards share an interest in medicine safety in toxicological, clinical or epidemiological terms. Back in the 1960s and 1970s, the pharmacologists and drug safety scientists who drove the creation of adverse drug reaction reporting systems in the UK and in France for instance were employed by hospitals and universities. Along with this primary occupation, they were sitting on technical advisory committees of health ministries. They were also closely involved with WHO activities and, for some of them, European committees. This is also illustrated by the fact that all three experts here are physicians by training, two of whom even started as hospital physicians before entering more directly the regulatory world as employees of a regulatory agency or as an industry safety officer. In these circumstances, the participation in European expert committees or industry working groups does not denote an attempt to accumulate resources and extend influence in regulation. It is in the continuity of their main occupation, the extension of a portfolio of activities that are traditionally connected to their occupation. Royer, a professor in pharmacology in a French university hospital, and one of the developers of the French and European spontaneous notification systems in the 1980s, thus explains:

Royer: Official institutions do not really deal with the scientific aspect of things. They take care of regulation. So, if we want to provide them with precise, reliable elements on which to base those rules, then we need to study those apart. And who does this better than academics? They are those who like doing this. They find topics themselves. At the last meeting of the International Society of Pharmacovigilance, there were I don't know how many hundreds of posters on subjects that were relevant for regulators [. . .]

Interviewer: Would administrations formulate particular requests to academics?

Royer: No, but you would feel the need, since we are inside official structures. Academics keep sitting on committees and administrations in different countries. So we feel the need, the problems, we transfer them into the scientific domain. We have all participated in committees or inquiry commissions, etc. And academics work for themselves, for the pleasure and for the publication of data, but as they do this they become experts. As they become experts, they are called to advise regulators at the national and international level. I have done this, I would almost say, as a normal thing. A function, a continuation of our professional function.[1]

The professional continuity between a job as academic or researcher and scientific adviser is harder to construct for food scientists, for whom the justification of participation in regulatory decision-making is not provided by the professionalisation of their discipline. To be sure, the evaluation of risks and of products has become almost occupationalised, with the institutionalisation of these processes and the creation of dedicated committees or organisations (independent regulatory agencies) as well as the increasing formalisation of the competences and processes of scientific advice (Groenewegen, 1991). The activity is also very frequent. Meetings may take place as often as once a month and require looking at large data dossiers, making product evaluation a time-consuming job involving personal dedication. On top of that, participation in these meetings is financially covered by the agency that organises them. While this money generally goes to the institute or university to which the research is affiliated, some also manage to keep the money, creating additional income that further motivates their involvement in expertise. The occupationalisation of these activities also means that scientists have access to them early on in their career, often through being co-opted by senior colleagues.

However, scientific advice is not professionalised in the sense that the possession of codified expertise and qualifications is not a necessity to gain access to these committees nor a protection against competitors. Each advisory committee, furthermore, calls on scientists from a variety of disciplines and research fields, depending on the agenda of the committee. For these reasons, the continuity with an academic job is not an obvious factor in the discourse or career of food safety scientists. They find themselves obliged to produce more complicated rationalisations to explain their involvement in scientific advice to government. They explain their entry into the world of official evaluation as a consequence of their developing a research specialism, and the coincidence between this specialism and the scientific information needs of public authorities. Advising governments is a topic-driven engagement.

In this sense, following his research interests and specialty, Nicolas (Chapter 6) gradually took on more posts in French and then European scientific advisory committees. However, he remains attached, subjectively and materially speaking, to the world of research:

> My position is special. I am a researcher, and I am paid for that. Therefore this research project is a collective one, a European project, through the 6th Framework Programme. And the third pillar, it is expertise. All of that holds together. These are not independent things. The knowledge gaps I see as an expert I can propose them to DG Research.[2] Well, I am not going to simply turn one random item of an expert meeting agenda into a research problem and experiment. But when there is an issue concerning food structure and allergenicity, it is obviously linked [to my research]. Therefore there is a connection at a more upstream level.[3]

Even though the skills and the subjects as well as positions and colleagues differentiate their scientific advice activities from research activities, most food expert scientists rationalise their activities for coherence, to the point that scientific advice is said to influence the formulation of research interests. Overall, advisory activities provide one more point of reference for researchers to test their work, formulate new hypotheses, orientate further research in the field and decide on topics where funding is likely to be available. Advisory activities are far from mere applied research. No discourse about the necessary involvement of the scientists in civic affairs is detected here. Only a certain form of pragmatism can be noticed, whereby scientists accept that their problematics and resources of study come from outside their academic field. The coherence in the different activities of Nicolas is ensured by defining a research programme (in this case on allergenicity) that can easily be aligned with regulatory needs and public discussions about food risks. It is also assured by a rather proactive attitude towards research funders and his own research institute, to make sure his themes are represented in their research agenda. The knowledge gaps identified during scientific advice and risk-assessment activities thus find their place in research programmes, in such a way that every piece of bibliographical work done as part of scientific advice, any data obtained or examined in scientific committees, become resources for his research.[4]

While Nicolas folds expertise back into research, other scientists do the opposite. They let experts problematics reframe their research in such a way that activities as experts may over time redefine disciplinary affiliations.[5] Patrick, one of the French experts involved in the discussions on PMM and head of a research centre on food and environmental risk-assessment methodologies (having more recently joined the WHO food safety services), similarly illustrates the influence on his research from issues encountered as a scientific adviser:

> As expert, it all started with food additives and the European Directive 95/2. That really was the starting point for my work. Following that, the link with my research interests basically trickles down a little bit from that. Over time, we started working more and more on the improvement of methodologies for the quantification of risks. Thence on tools, to state was the risk is.[6]

These three examples show that expert activities impose on food safety scientists some transformation of their work, which can be contrasted with the continuity between professional and regulatory activities put forward by pharmacologists. In both cases however, the necessity to generate knowledge on new substances or on new adverse effects, possibly also the development of new methodologies and tests, sparks a reprofessionalisation of scientists. It generates involvement in activities beyond teaching and research as they associate their work more readily with the accomplishment of a service of public utility.

The case of HACCP specialists is an intermediary one. Food microbiology has traditionally been a discipline that came in support of the food industry, be it in product development and innovation or in matters of food hygiene and quality control. Furthermore, food microbiology and veterinary medicine are areas of expertise that are recognised, by law, as prerequisites in many countries to be a veterinary and food hygiene inspector. Jean-Claude, by teaching in a national veterinary school, is officially employed by the French ministry of agriculture which supervises these schools. He trains inspectors and civil servants working in administrative services responsible for agriculture, food control and food safety. It was very natural for him to be called on to contribute to the elaboration of the ministry's policy concerning food quality, when the issue surfaced in the early 1980s. The same goes for Michael and his move from a national laboratory to Nestlé and in parallel his increasing participation in advice to government and the WHO. Altogether, it explains the orientation of the ICMSF to contribute to the development of international standards and policies and the absence of any friction when they move between academia and research, industry, and giving advice to national and international standard-setting or policy-making bodies.

A characteristic of these scientific experts is therefore that they belong to the elite of their research area and become involved in a more or less occupationalised role of risk evaluator, which makes them move away from their initial position (e.g. as hospital physician) or complement their main job (e.g. as academic or researcher) to take a more hands-on approach to the evaluation of adverse health situations and through this, connect to either of these spheres of government or industry with which they originally had less contact. This occupation extends through participation in a diversity of advisory committees, more or less in continuity with this engagement in the job of assessing risks. These experts are experts because they are concretely embedded in regulatory activities.

TRANSNATIONALISATION AND CIRCULATION

These actors combine a second element with multi-positionality. Common to all scientists in the three cases is the fact that they are also crossing the boundary of their originating country to work with international standard-setting and policy-making bodies. This transnationalisation is true for all of the missions, consultations, committees or groups they participated in, whether academic, professional, industrial or intergovernmental. Gabriel, as he moved from a position of hospital physician to a position in a large multinational company, quickly started participating in the relevant working group of the European pharmaceutical trade association. In conjunction with this, and as early as in the mid-1980s, he was involved with the CIOMS in popularising industry activities and processes of pharmacovigilance. It is also noticeable that he started to participate in meetings of the European Association for Clinical Pharmacology and Therapeutics as well as the Société Française de Pharmacologie. Over time, his reputation as an expert in industrial pharmacovigilance, confirmed by his various memberships and activities, led him to take a bigger role in meetings of the EMEA when the latter was created, as well as in the ICH. In other words, the career and activities of Gabriel became more transnational as pharmaceutical regulation itself became Europeanised, and later internationalised with the ICH process. The trajectories of Paul and of Patrick Waller also confirm this phenomenon, with them both extending their activity of assessor at the MCA into participation in European committees. In the case of HACCP experts, the transnational movement is further reinforced by their belonging to the ICMSF – from the early 1980s for Michael, and early 1990s as concerns Jean-Claude. The novel food experts all have in common that they have moved from participation in national scientific advisory committees to European ones, and to have been early members of the influential Scientific Committee for Food of the European Commission. In all cases, they have thus become transnational scientific experts as the European Union asserted its pharmaceutical and food policies in the 1990s and 2000s and as international standard-setting bodies gained in importance.

The impression that there is a distinct 'transnational' scale of activities comes from the gradual accumulation of missions, reports and committee memberships by experts over time, to the point of never fully retiring from them and returning 'home', even after the end of their main professional activity. But rather than being a separate scale of activity, the transnational should be understood dynamically because the starting point for access to these missions and positions is a combination of occupational and regulatory activities that is created back home. Circulation starts within a national context and then extends in such a way as to form a unique space of circulation with much fainter geographical boundaries. In this way, Gabriel moves around between the CIOMS, the EFPIA working groups and ICH meetings. Paul's membership of the CPMP coincides with participation in CIOMS meetings and a position as EU representative in the ICH working group on drug safety. These activities led him, as an extension of these roles, to join the European Commission in 2004 (pursuing his work on pharmacovigilance) before returning to regulatory activities at the EMA. The turn of the 1990s for Jean-Claude coincided with joining the ICMSF. In 1997, he abandoned his professorship to be recruited as principal administrator at the European Commission and later as a head of unit at the FAO. Michael moved on the same scale between the ILSI, codex meetings, and advisory activities to the WHO. Nicolas, Patrick and John all followed up on their membership of the Scientific Committee for Food with participation in EFSA activities and the ILSI, but also the JECFA (for Patrick), the OECD and the FAO (for Nicolas).

Quite strikingly, this accumulation of roles transnationally had an effect on their careers since several of them moved further away from science and evaluation towards a policy role as exemplified by Paul (joining the European Commission and the EMA), Jean-Claude (the FAO) but also Patrick (who joined a policy unit of the WHO in 2009). Figure 7.2 represents this, showing the three typical trajectories followed by 24 scientists involved in the development of pharmacovigilance planning, post-marketing monitoring or HACCP.

The vertical axis concerns the main interests of scientists going from particular substantial matters of products or risks at one end to regulatory concepts and standards at the other. The horizontal axis shows the affiliation of scientists to an academic or research organisation, including

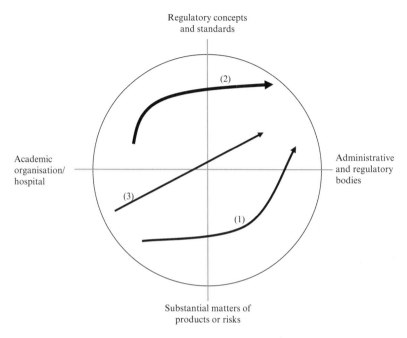

*Figure 7.2   Typical professional trajectories of scientists in invisible
colleges in the case of PVP, HACCP and PMM*

university hospitals, or to administrative and regulatory bodies. The
centre space refers to participation in scientific advisory committees, in
many ways a point of equilibrium and first exposure to regulatory activities for many scientists. Being a member of a committee is a step towards
a job as regulator and marks an evolution in different research interests of
a more regulatory nature as seen above.

In all cases, a rise to more administrative functions and interests in
regulation can be observed, concomitant with increasing distance from
experimental research. The first trajectory is typical of specialists of
product safety, particularly of medicines. As they belong to the medical
profession, prominent positions in hospitals and universities as well as the
pharmaceutical industry are accessible. Participation in scientific committees and employment by public administrations or agencies may thus take
place rapidly, with a corresponding increase in the importance of concerns for the criteria and protocols of medicine safety, beyond particular
cases. The second trajectory is symmetrical to the first one. It is typical of
microbiologists who do technical research into conditions of production

or particular bacteria from an interest in quality assurance systems such as HACCP and, later, in risk analysis. As they deal with these regulatory or political themes, they also take on administrative or regulatory commissions and positions. The third trajectory is followed by scientists in all domains when they turn their research specialty into an expert specialty by being members of scientific advisory committees that deal with their topics of research and discipline. The noticeable fact in this case is that participation in a committee often leads to involvement in other committees. Scientists seldom cease to act as advisers. They stay on the same committee for several terms (unless this is forbidden), and move to another organisation in the same country or at the European or international level. As agencies lack but need advisers, the same people are asked to participate. Another phenomenon is that scientists may drift away from research as they spend more time on scientific advice, and will thus be drawn to more administrative functions. A career as scientific adviser rarely ends before retirement since one commission leads to another. Only a few scientists among those studied in this research stopped acting as experts voluntarily (i.e. without being obliged to step down because of compulsory retirement age for example) to return to teaching or research activities. This trajectory is more than a journey through a professional field. It is also an evolution across a research field as hinted in Figure 7.2, from researching particular adverse events to researching typified adverse events and theorising interventions. This means that defining the list of risks to attend to becomes more important than investigating one in particular. Establishing and standardising methodologies of risk evaluation and risk control take precedence over their mere application.

Concomitantly with this establishment at the transnational scale, circulation between the different spheres of regulatory activity also appears easier, and the neutral hat of 'expert' seems to fully play its part in allowing scientists to freely move between professional, scientific, industrial, intergovernmental and supranational sites – an important factor in the ability to pursue the development of a concept and its standardisation. Gabriel, although from the industry, makes it clear that at a transnational level he does appear more often as a detached 'expert', standing for his knowledge and experience rather than for corporate interests. The development of the science of pharmacovigilance gives a coherent motive to transnational activities:

> I have four sorts of activities. The first one concerns the database, with individual notifications of adverse drug reactions. The second concerns the more synthetical part, which consists in detecting signals of adverse drug reactions by using different types of methodologies on the dataset. This I do in collaboration

with global product managers. For each of the products, we regularly review the data to see what works and what does not. The third element in my professional activity is crisis management – a pervasive activity, which takes all of my time these days. And lastly there is my fourth function, or activity (which is I have to say very time consuming but highly interesting), that is to represent my company in a number of working groups or committees, with regulatory authorities and professional bodies. Or in the framework of more general reflections on pharmacovigilance: on operational problems in pharmacovigilance, those that are posed to pharmacovigilance people on a daily basis in professional terms, such as signal detection, signal evaluation, clinical trials, how to design trials, how to anticipate safety issues, etc. I am going to the CIOMS for that. I am sitting there as an expert: I am not representing my company there. And I do all of this to contribute to advancing the discipline. It is the motivating aspect in my activity I should say.[7]

In Gabriel's depiction, the advancement of the discipline represents a sort of social and political function, that of dealing with the problems of a profession that itself has a social role – one of advancing safety. Working on methodologies and protocols for safety evaluation and management is a logic that he can implement in the roles where he represents his company and where he acts by virtue of his experience and knowledge 'as an expert'. However, being an expert does not restrict his activity within the CIOMS even if it contrasts with his activities in the ICH as a formal representative of the industry. Gabriel's work in the ICH is still an extension of his activities in the CIOMS; he is carrying within the ICH the concepts he helped develop within the CIOMS.

This is only an example, but it is further compounded by the many regulatory roles pursued by the other scientists in similar ways. It shows that the transnationalisation of their careers coincides with a greater facility to work across different spheres to further mediate them through conceptual work.

DISTRIBUTED STRATEGISM

The above quote from Gabriel points to a third differentiating characteristic of invisible colleges, which is their relation to the action of interstitial organisations such as the CIOMS, the ILSI or the ICMSF. Invisible colleges and these organisations dovetail. The former are constituted and convened by the latter who give scientists an opportunity to act strategically towards transnational regulation, with minimum commitment and need for a public role. The latter, in return, obviously gain authority and credibility, and ultimately much sought-after influence over international standards.

Interstitial organisations are of various types (professional organisation, trade association, think-tank). What they have in common is that they are located between several spheres – government, industry and science – and establish relationships between the organisations in these different spheres. These organisations set up working groups and recruit individual scientists who work with governments, international organisations, civil society or industry. They indirectly link all these other organisations and improve coordination. They are essential for scientists of invisible expert colleges to introduce collegiality and foster data exchange as well as concepts for evaluation. Membership of interstitial organisations reveals the existence of groups of experts who are motivated to forge links with the regulatory domain, channel discussions on particular risk experience and articulate shared regulatory concepts. Their participation in the activities of these organisations goes beyond being members of a working group. They willingly chair these groups and cooperate with the secretariat of the organisation to define its mid- and long-term agenda.

The CIOMS, for instance, is a forum for the 'biomedical community'[8] to discuss issues that no other governmental or intergovernmental organisation really owns as an issue of policy or regulation. It is recognised as a useful forum of discussion by the WHO. All its activities are funded by its members and meeting participants. The CIOMS has been used by the pharmaceutical industry for progress on drug safety and pharmacovigilance, at a time when no industry body or government offered the means for establishing policies about it. In the early 1980s, meetings were organised under the aegis of the CIOMS to reflect on the harmonisation of company practice for post-marketing drug safety and, more crucially, to create an agreed-upon format for reporting adverse drug reactions. This first series of CIOMS meetings on pharmacovigilance was initiated by industry and academic pharmacologists who could use this conference as an appropriate channel for this enterprise: one that could allow for both an open and disinterested discussion around standards of information transmission, and convey its intention to actually contribute to governmental drug control activities. While certain governments or the European Commission have at times dismissed the legitimacy of the CIOMS because of its lack of representation, distance from organised interests is also a factor in its credibility. In particular, the dissemination of its standards critically depends on its ability to embody a form of neutrality and universality.

Both Gabriel and Paul have affinities with the CIOMS, having been involved in the developmental odyssey of international pharmacovigilance since the 1980s. They recall the rapid progress achieved at CIOMS meetings on difficult topics such as harmonisation of pharmacovigilance dictionaries and reporting formats to ensure that data about side-effects coming from different countries may be exchanged and accumulated. Royer and Gabriel have been involved in CIOMS meetings regularly over several decades. They also actively defend its products and utility in general as a provider of regulatory ideas and standards, particularly for governmental authorities. The possibility to work in the CIOMS in their own name, leaving their affiliations in the background, was decisive in dissasociating the concept of planning from a promotion of pharmacovigilance by its professionals to representing it, on the contrary, as a solution to a common problem of inefficient risk minimisation.

A second organisation that plays an important role is the ICMSF. Although closer to a learned society or a small professional club, the ICMSF has nonetheless acted on several occasions in similar ways to the CIOMS: as a forum for the advancement of topics that were felt to be too specialised and too promising to be dealt with by government officials and through intergovernmental negotiations. The ICMSF has been trusted for its credibility as a provider of regulatory ideas for its members – a closed group of internationally renowned co-opted microbiologists. Equally, ICMSF members put forward innovative, ground-breaking and influential ideas to governments and international organisations in the area. The scientists are individually committed to the organisation, ensuring it maintains its high credibility through a strict policy of co-opting the very best food microbiologists.

The third case is that of the ILSI, a foundation with branches in the USA, Europe and Asia. Although it is funded by the agro-food industry, it has legitimacy in the eyes of governments as a unique forum of scientific specialists on particular classes of products and for the discussion of the regulation of these products. The same dual attachment of scientists to the ILSI can be observed as is the case between drug safety experts and the CIOMS. Food safety scientists are independent of the ILSI as the latter's resources are unnecessary for their daily work and the publication of their research. These scientists are not paid by the ILSI. Their participation at ILSI meetings of academic scientists is covered by the ILSI or by companies for industry scientists. The work of the ILSI is valued by the scientists who work with it. Apart from the fact that it is funded by the industry, they consider that the ILSI provides a unique forum for governments, industry and academic scientists to meet and exchange their views on products, risks and evaluative methodologies. The ILSI also increases coordination among different teams of scientists in the same field. It sometimes carries out collective EU-funded research projects. Another advantage of the ILSI is its international coverage. In line with this discourse, the different food safety scientists involved in PMM or HACCP have also been in contact with the ILSI on more than one occasion. The links with the ILSI however are intermittent.

Transnational interstitial organisations are lean organisations that leave scientists free to decide on topics to work on. They are vessels for the concepts that scientists want to develop and appear to have no other agenda than that of their many members. They have minimal staff and skeletal structures and are not tied to the production of a particular output. Even when new concepts are produced and new standards shaped, the typical discourse is that regulators are free to take up the new concept and that lobbying is outside their remit. Delegating the gestation of concepts and the lobbying work to scientists, these organisations maintain their reputation of independence and disinterestedness. In return, scientists have the opportunity to contribute to regulatory debates without exceeding their role as scientist or breaking these very norms of independence and communism that still regulate part of scientific life and its image.

These organisations also use and uphold scientists' collective ideal of working by consensus. Since participants do not represent interests outside the meeting it makes it easier to overcome conflicts among actors with different positions on the rules applicable to pharmaceutical or food products, as is often the case transatlantically or even within Europe. In the meetings of such organisations, people's affiliations are tacit. Being present, these affiliations are known and can be worked around to link together interests and experiences. At the same time, holding them tacit – or invisible – is a way for everyone and for the organisation as a whole to demonstrate that the logic of forming links has primacy over others. Not making affiliations known also pushes actors to take risks: it means they will not be blamed or held accountable for what they do together back in the area of their activity as researcher or professional. Invisible colleges are not tied to a standard output that they have to produce at all costs. They work from what they have, and if they have something to build on. Invisible colleges, furthermore, are not part of regulatory competition. They are not threatened to be overtaken by other groups, and are not governed by obligations of productivity and efficiency.

Interstitial organisations play a crucial role, which is to allow scientists to engage in strategic behaviour in a low-key mode, through weak ties with fellow scientists, in such a way that it does not fully redefine them as regulatory activists but on the contrary helps them to maintain their own interstitial position and circulation. Outside particular episodes of standardisation, the social ties between scientists are generally weak. They are often geographically distant. They are also based in different sorts of organisations, as highlighted above. They do not necessarily belong to the same schools or research traditions. Interactions between them may accelerate during periods of debate around a particular standard, but then slow down or end once it is adopted. Even during the phases when publications, reports and official meetings multiply, interactions remain intermittent and certainly do not occur daily, weekly or monthly. They are also provisional as they rarely go past the fleshing out of the concept and adoption of a standard. No two concepts are elaborated by the same college, comprising the same scientists. There may be an overlap between the membership of two colleges, in particular when two concepts are interlinked and individual members jump from being experts on one to experts on another, as was the case with HACCP and microbiological risk analysis.

Interactions are therefore weak, but maintained by interstitial organisations. Interstitial organisations create a distinctive mix of formal and informal, face-to-face and distant relationships among scientists, which is conducive to solidarity. The relationships that are forged through these organisations are formal; that is, they are established in the framework of an organised project, report or publication. These temporary structures of knowledge production help the shift from the decentralised and often tacit circulation of a set of ideas to the explicit and formal production of a stand-ardised concept. Interstitial organisations allow a shift to take place, from the plurality of experience of interventions to the uniqueness and coher-ence of formalised knowledge and standards. The formal aspects of these relationships are in the project-based contracts with the ILSI that result in the publication of papers and public events. But these relationships often alter to a more informal mode while collaborations may continue in a less organised and instrumental form. Collaborations develop through email and exchanges of papers as much as through physical meetings. Over time, the relationships sponsored by these organisations become independent of them. Fellow scientists who meet through these organisations integrate the habitual networking of scientists. Although their interactions through these hubs are sporadic, scientists do value these organisations for opening up relationships with other scientists on shared research topics. The low frequency of meetings is important for the collective endeavour not to turn into a separate enterprise that would absorb all the scientists' efforts and redefine them as policy entrepreneurs. But there are enough meetings, and on topics that matter to the delegates, for the interactions to be enriched by a de facto solidarity in the success of the standard.

This mode of interaction allows scientists to engage in standardising work without giving up their scientific ambitions and their image of being independent experts. During these interactions, innovations in medical and food technologies are discussed, unseen adverse situations assessed, and new concepts considered. Scientists work with other prominent scientists in their field on topics of regulatory and often public interest, offering an opportunity to make a decisive contribution. When they meet, it is to explore a new research and regulatory front: all three concepts of PVP, HACCP and PMM were highly innovative when the CIOMS, ILSI or ICMSF started work on them. The scientists brought together by these institutions were conscious of how few of them there were, but also how innovative their concepts were. One of the French specialists of PMM and exposure to chemical contaminants notes that there are only five such specialists globally. The HACCP specialists pride themselves on being part of a small mafia of experts who are repeatedly called on by governments and international organisations in the field to contribute to the discussion on the matter. They emphasise readily that they belong to a small world with an even smaller number of specialists involved in regulatory pharmacovigilance at senior level or that, out of the large population of people who experiment with HACCP and publish on it, few would be called on by international organisations to contribute to the elaboration of the standard.

The strategism of the scientists who interact with interstitial organisations to convene invisible colleges should in a sense not be understimated. In many ways, these experts are political as much as scientific experts. Many of the experts consulted by Waller and Evans had enough prescience of the evolution of the industry's competence, internal organisation and strategic objectives towards post-marketing safety, to be able to guess that a sufficiently large part of the industry would support a proposal to set up a PVP guideline. For instance, they anticipated that the flourishing global analytical and training services industry would quickly hone in on the necessary activities – conferences, training – to spread risk-management competence and schemes in the industry. All PVP proponents working within industry or in regulatory agencies know that risk-management practices are able to spread quickly through an industry as soon as an emergent group has pledged to adopt the practice. As participants in CIOMS meetings, having witnessed the engagement of industry specialists in the development of new and sometimes costly standards, they shared this understanding of public action dynamics within the industry. The whole HACCP idea is based on the belief that there is a convergence of objectives from the industry (gaining the right of self-regulation on food hygiene and softening public controls), national administrations[9] and the WHO. The concept was born from a company's publicly inspired and funded project (Pillsbury's work for NASA). Much of the ICMSF experts' involvement in the development of HACCP follows from industry's demands to have a benchmark of food safety assurance. Both Jean-Claude and Michael for instance testify that food companies proactively participated in conferences on HACCP and required them to participate in the development of this tool as well as its application through consulting missions. Thus Jean-Claude sees himself and many of his colleagues as anything but 'stubborn leftist academics' who refuse to speak to the industry. This also applies to PMM experts who all cooperate closely to some degree with the industry as product evaluator, researcher or consultant. Their involvement in the ILSI is underpinned by the common comprehension that aside from profit objectives, industry experts need to engage with academics and scientific advisers to government to respond to public concerns. Many of them had had previous experience of this 'honest' commitment (to use a word employed by John) in schemes of product surveillance as consultants on such projects (as is the case of Nicolas and Unilever) or by assessing their experimental surveillance schemes (as John did in 1983 for RHM).

Scientific experts before they formally work together on a concept are commonly aware of two points. The first is a sense of who has influence in the regulatory system. References to various important personnel who had a particular bearing on regulatory debates fill their stories and comments that show how lucid they were about the impact of unofficial and little known bodies such as the Heads of Agencies group, CIOMS, ILSI, and so on. The second is an understanding of whether or not there is room for agreement between industrial, professional, governmental and intergovernmental actors. In each of the examples in this book, the concepts have come a long way to be recognised as the internationally accepted standards and natural language on the types of practices and interventions that constitute public regulation. It is of course easy with hindsight, to argue that PVP, HACCP or PMM are the best possible formulations for methods of surveillance preparation, the practice of surveillance or food quality assurance. Seen from the other end, the successful trajectory of regulatory concepts is a rather puzzling phenomenon. The resources that these scientists invested to develop and promote the concepts show that they anticipated wide acceptance and effective framing of regulatory practice. This does not just denote a capacity to influence other regulatory actors. It implies a sense of what could work, how a concept could match these actors' interests, expectations and existing experience of adverse situations and ways of addressing them. At several points, the scientists involved in the above stories showed that they believed their concepts commanded consensus. Their becoming a formal standard was evidence, in light of the mileage these concepts gained across the whole of the regulatory system in the discourses of a variety of actors. This is evident from Waller's appreciation that 'lots of people had similar sort of ideas' about the overhauling of pharmacovigilance, to Paul's sense that 'he was pushing an open door' and Jean-Claude's appreciation that HACCP 'was in the air' in the early 1980s in the USA.

This shows that interstitial organisations allow scientists to build on the potential for strategic action towards transnational regulation. Invisible colleges do not meet altogether at one and the same time. They are represented by small pockets of two, three, four scientists. But these few take particular responsibility towards enrolling others in working groups, meetings, expert consultations, and so on. Scientists use these organisations to gather fellow scientists together to work on particular concepts, and their mutual relationships are consolidated in return by their activities in these organisations, which allow them to achieve their ambitions to contribute to regulation. Interstitial organisations help scientists to distribute their strategic behaviour across a collective, and to better distance themselves from the behaviour of a lobbyist or policy entrepreneur that does not fit with the label of expert. By collectivising these strategies, convening others in organisations that in appearance do not defend any particular industrial or governmental interest, they maintain an image of neutrality and brokers of consensus. The pursuit of their own research and professional interests becomes intertwined with this ambition to include all actors in a common circuit of standardisation.

CONCLUSION

The scientists who take part in these loose collectives have several things in common, which define what a scientific expert is in these stories of standardisation: that of being part of the elite of a research field, an initial occupation that favours circulation towards more roles, and an intellectual and professional trajectory towards a more central position and interest in regulation, concomitant with a transnationalisation of their career, and their participation in the life of open collectives and interstitial organisations that allows the latter to be kept together. I have argued that invisible colleges should not be dissociated from, and are indeed observable through, the activity of these organisations that, by gathering highly mobile and multi-professional scientific experts and banking on their knowledge, shape the transnational circuits by means of which concepts are standardised. They would not achieve this without the contribution of a few scientists who are distinct in enjoying work on regulatory issues, become quasi-regulators as they progress in their career, and have a clear ability to confer with and mobilise their colleagues.

Scientific expertise in general and evaluative, standardising sciences in particular appear to be more complex than what is assumed when an epistemic community is being credited for the formation of an international regulatory regime or when neo-institutionalists trace the origin of world-models to 'the highly legitimated cohesion of this professional group, bounded by a common ethos, [that] helps make science an influential organized force' (Drori et al., 2003: 291). But rather than groups and a common ethos, this chapter found loose and open collectives. The interactions within these collectives are irregular, provisional, and unstable. Its members are linked through weak rather than strong ties. They meet face to face in smaller numbers, but these micro-groups are linked to a broader circuit by previous collaborations or publications. No two scientists really have the same professional trajectory and experience, and it is hard to find any underlying common interest or ethos that could justify going beyond an ephemeral participation in standard-setting committees and working groups. What has been exposed here indicates that these episodes serve in hindsight, as the discourses of these scientists often show, to argue that a group, 'mafia' or 'club' is behind the adoption of a standard. This is obviously a reconstruction, given that standards are rooted in standardised regulatory concepts that no one in particular can be said to own. These discourses however illustrate how much value is placed by these scientists on these specific episodes, which in hindight give meaning to their circulation as well as their assembling in interstitial spaces in which they can fulfil a political ambition to rationalise regulatory practice. The final chapter is the occasion to revisit the differences between the three cases to find an explanation for why certain invisible colleges seem to go further than others in this rationalising ambition.

NOTES

1. Interview with René-Jean Royer, 16 January 2004 in Brussels.
2. The European Commission Directorate General for Research, in charge of the elaboration and running of framework programmes for research and technological development.
3. Interview with Nicolas, 10 September 2004 in Paris. Nicolas uses the generic term 'expert' to speak of his activities of scientific advice.
4. Anecdotal evidence also suggests that per diems and expenses reimbursements constitute a potentially interesting parallel income, especially for those scientists who combine memberships in several committees.
5. A British ecotoxicologist, adviser to DEFRA, former member of the European Commission Scientific Committee on Plants and then member of an EFSA panel, thus recounts: 'My undergraduate degree was environmental biology. My postgraduate, my PhD was in behavioural ecology. Of birds. And I then joined the ministry of agriculture, the same job that I now have. For a long time, if you'd asked me in the 80s and 90s what my discipline was, I would have said eco-toxicology. And if you ask me now, I would say I am a risk analyst or something like that. Because I broadened my area by getting into quantitative methods first, and by extending those to other types of risks.' Interview with Andy Hart, 7 October 2004 in Brussels.

6. Interview with Patrick, 16 June 2004 in Paris. Directive 95/2 mandated a re-evaluation of all food additives authorised in Europe.
7. Interview with Gabriel, 30 January 2004 in Paris.
8. From the CIOMS website, www.cioms.ch/about/frame_about.htm, accessed 20 June 2009.
9. Those in charge of consumer protection in particular, historically coming second to agriculture ministries for food hygiene and food safety controls.

8. Scientists, standardisation and regulatory change: the emergent action of invisible colleges

Many scientists are in the business of evaluating risks. Many also demonstrate a certain degree of polyvalence and multi-professionalism in working with bureaucracies, industries and international standard-setting or policy-making bodies. This mobility and circulation is not the preserve of the scientists followed here. Settling down in more official regulatory or policy functions is rare though, as is the fact of being consistently involved in a circuit of exchange of experience and ideas about risks and their control. Being involved in standard-setting projects is rarer still. Not all scientists can hope to have the impact that the three regulatory concepts of pharmacovigilance planning, HACCP and post-market monitoring had. There are also some differences between the three cases in that respect. This chapter revisits them, shedding more light on the changes that they brought about in the control of products and adverse events. It does so to solve one final question: are invisible colleges political actors who seek to achieve defined effects on regimes and markets? How much do they seek and take responsibility for imposing defined regulatory change on other actors of the regulatory domain, as opposed to articulating and enabling change that is expected and desired by the latter?

The impact of regulatory concepts on activities of control can be assessed at the level of systems of control. Risk regulation is an interplay of tools and actors – sometimes also called 'regimes' (Hood et al., 2001). As is clear from all three cases considered here, regulatory concepts do institute such systems. They define criteria of control, hence criteria of overall success or failure of regulation. They also define the mode of coordination of businesses and regulatory agencies through common protocols of monitoring, planning, analysis and correction of contaminations, and so on – thus constructing possible responses to failures. PVP established cooperation between different services and teams from the product manufacturer. These cover marketing, internal product sponsors following development and testing, and pharmacovigilance. Outside manufacturers, PVP includes specialised information and consumer survey services, and experts reviewing plans on behalf of regulatory agencies that monitor its implementation and alert the public and professionals when necessary. HACCP also creates a range of new identities of intervening actors and relations between them, from the company's HACCP team and officer, consultants who help establish the plans, and certification companies, to risk assessors who review the data generated by this monitoring process and set targets in the industry or in regulatory bodies. Similarly, the post-marketing monitoring of novel foods rests on a chain of actors. They extend from the development and testing services of the manufacturing company, specialised surveys and polling services for interviewing consumers and food distributors, and scientific advisers and staff assisting them in regulatory agencies, to the bureaucrats and politicians who make regulatory decisions about the products. So, do scientists engage in evaluative and standardising actions because they want to implement change in the system of control and respond to its failures? Do they assume responsibility for the overall system?

If so, it would mean that they are full political actors, with certain purposes and a dose of legitimacy accruing from their responsiveness to adverse events and to the regulatory failures that they sometimes represent. This would be an interesting characterisation in the light of the argument of world culture theorists that standardisation as a form of rationalisation does not make someone an 'actor'. Quite the contrary, invisible colleges as studied here would fall in the category of what Meyer calls 'rationalized others': those 'social elements such as the sciences and professions (for which the term 'actor' hardly seems appropriate) that give advice to nation-state and other actors about their true and responsible natures, purposes, technologies, and so on' (Meyer et al., 1997: 162), without taking much of the responsibility for their effects (Meyer, 1994). Although insightful, this claim of Meyer and colleagues is not entirely accurate. At least it does not apply in all three cases here. It is salient in the case of HACCP and PMM, where scientists appear to have standardised the control of foods without changing much of the system of control, its boundaries or responsibilities – including their own. HACCP rose to the status of a mandatory system in about 20 years. As a generic concept, it was implemented in various different ways through multiple national or sub-sectoral regimes to better fit with the areas of food production for which such abstract managerial principles seemed too prescriptive. Failures in implementation were unforeseen by HACCP proponents. Regarding PMM, the integration of a mention of the concept in an article of the future European Regulation for novel foods is still under discussion, nearly 30 years after the first PMM-type experiments. It has a very restrictive scope and refrains from precise codifying of the procedure for requesting and interpreting these studies.

In a sense, the concept was designed to ensure minimal disturbance on the regulation of novel foods. But the argument about the absence of responsiveness and responsibility of invisible colleges seems less relevant in the case of PVP. It took only a few years before PVP-type concepts emerging from professional associations or regulatory agencies were channelled transnationally into a common notion of pharmacovigilance specification via the CIOMS and MCA, progressing quickly thereafter to become an ICH and EMEA guideline. Detailed practices and criteria of control were prescribed in these guidelines, and the changes corresponded to the designs emerging in the college. There was an element of purposive action in Paul and Gabriel's enterprise of pushing PVP at the ICH, not least to strengthen pharmacovigilance in line with the risk-management agenda of leading regulators and politicians.

Invisible colleges may or may not act intentionally and responsibly towards the designing of a regime to give coherence to the actors and tools that compose it. PVP proponents did work to change regulation as a whole and aimed to produce particular effects on this regime, following their professional objectives of minimising unexpected adverse events and drug-induced morbidity, but the scientists involved in the two other cases did not. The comparison reveals that, first, the degree of scientists' engagement with the whole system of control is coherent with the coverage and composition of invisible colleges. In the case of PVP, it had a wider reach and channelled more experience of adverse events owing to the professional responsibility of drug safety evaluators in the control of adverse drug reactions. Second, invisible colleges and regulatory concepts emerged from regulatory domains. They are a reflection of the constitution of these domains, their boundaries, distribution of roles and responsibilities, and outcomes. Evaluators have a collective capacity to change systems of control if they have a responsibility for them and a likelihood of involvement in constructing – and remedying – their failures. This means that invisible colleges are emergent from domains aiming for a high level of control over products and adverse events and conversely, are more susceptible to failure.

CONSTRUCTIVE FAILURES

A regulatory domain can be more or less bounded and integrated. The controls in place may or may not form a system, depending on the precision and acceptance of the criteria of intervention and of the protocols that design roles and operations of control. All three regulatory concepts examined here seem to shape such a system, albeit to different degrees, with PVP doing it more than the other two.

PVP is both precise and large in scope. First, it proceeds from a concept that was initially inspired by a handful of cases, starting with cerivastatin. But the lessons drawn from this case were rapidly put in line with broader knowledge of the limitations of pharmacovigilance, mainly of spontaneous notification systems, to forge a concept of pharmacovigilance specification with great validity. It applies to all new medicinal products, that is, to all substances qualified as new and authorised in the market. The EU legislation and guidelines that finally define pharmacovigilance or risk-management planning make it clear that it concerns all new as well as older products if a new safety issue arises. As illustrated in Chapter 4 these issues are described in a long list that risk-management plans must include. Second, PVP is quite precise or 'finite' (MacKenzie, 2008) in its prescriptions. The EMEA guideline in particular leaves little to the imagination. It includes a description of when the pharmacovigilance services of a company should come into contact with regulatory agencies. In this respect, PVP reconfigured the criteria and protocols of control in unequivocal ways. Risk-management planning has become a reality in terms of organisational processes in the industry and just as quickly in regulatory agencies. Risk-management plans are now routinely produced by the industry. Universality and prescriptiveness thus add to the swiftness with which these new concepts and attached standards appear and are implemented transnationally.

This shows that the intentions of its proponents were clear and precise, and were so before PVP took the shape of a guideline, when Waller and his team were reflecting on the content and implications of pharmacovigilance specification. It was understood in this invisible college that the introduction of pharmaceutical risk management in the industry and in regulatory agencies would require associating the actors of product development, and their culture of innovation in positive pharmacological effects, with processes of reporting of incidents and their side-effects. It would also require orchestrating close relations between businesses and regulatory agencies. To design these changes, PVP experts in the invisible college could capitalise on their experience of company-wide processes of industrial pharmacovigilance that were inaugurated in the 1980s with such tools as the periodic safety update report or the mandatory nomination in each company of a person responsible for pharmacovigilance. Both rules led pharmaceutical companies to maintain large systems of analysis and reporting of warnings of adverse drug reactions as well as safety databases. The concept of planning links pharmacovigilance and pre-marketing personnel who all have to contribute through these devices and systems of information. These reorganisations were anticipated by Gabriel given his function as the person responsible for pharmacovigilance and his knowledge of industrial pharmacovigilance (see Chapter 4, p. 99). PVP thus embraces a system of controlling medicines through two symmetrical tools of marketing authorisation and post-marketing surveillance. It preserves the coherence of the pharmaceutical regulation regime and brings together industries and regulatory agencies, and pre- and post-marketing safety services. This capacity to design and apply changes to the system of control can be illustrated in several other ways.

First, it is apparent that the evaluations of the invisible college concern simultaneously adverse drug reactions and the limitations of the marketing authorisation tool in anticipating or detecting them. Mandatory marketing authorisation delimits the boundary of medicine regulation. It defines which substances are in the domain and which are not. It also defines whether a company is a pharmaceutical company, since it is only really one if it holds a marketing authorisation for a product. The strengthening of pre-marketing tests and trials over the last decades has led to the rejection of more and more substances outside the space of substances considered for use in therapy and, in this quality, to be monitored. Since the medical profession created these tests and trials, it was instrumental in shaping the domain.[1] Medical professionals have thus appropriated the domain as an overall jurisdiction (Abbott, 1988). They even make politicians and bureaucrats redundant as gatekeepers of the domain. Pharmaceutical regulation in large part is operationally delegated to them, as shown in the autonomy acquired by scientific committees of the EMA for instance to make de facto decisions on marketing authorisations.[2] Pharmacovigilance and PVP continue this construction of the domain by the medical profession, building on the limits of marketing authorisation and revising the decisions that were made.

As shown in Chapter 3, the post-marketing safety specialists emphasise the limits of marketing authorisation, defining major unexpected adverse drug reactions as failures of this control. Pharmacovigilance is founded on the notion that this tool fails, allowing substances to enter the market that should not be there, or not in the conditions that were specified after toxicological and clinical testing. Medicine evaluators have leverage on a system of control because the risks that they typify – of certain classes of products like statins, the level of drug-induced morbidity, and so on – are inevitably a failure of the marketing authorisation tool. The evaluation of regulatory controls does not stop with marketing authorisation. All other controls introduced since then face similar scrutiny. Waller and his concern for the experimentation of PVP after realising the enormous resources invested in PSURs with little effect on drug iatrogeny is just one example. PSURs are known to be too time-consuming, undynamic in providing a picture of side-effects and insufficiently synthesised. This explains why PSURs did not prevent untimely drug market withdrawals for cerivastatin and other products. However, the deficiences of PSURs provided the catalyst for PVP.

Second, drug safety experts showed their interest in the control system of medicine safety by evaluating, at least rhetorically, its outcomes. All scientists involved in the invisible college, from Waller to Gabriel, could claim that they were acting to protect public health thanks to the virtuous circle of activities that PVP was supposed to create. The epidemiological influence plays out quite strongly here, providing these expert drug safety evaluators with a critical point of reference – the level of drug-induced mortality and morbidity. The discussion by Waller and Evans on the lack of coherence and drive in the development of pharmacovigilance, and the argument made by Waller that periodic safety update reports do not bring about as great a public health benefit as they should have, are the clearest illustration of this. The observations of Gabriel on what the PVP guideline would bring about, as much as Paul's appreciation of the lack of robustness of pharmacovigilance at the European level, are other testimonies of this integrated appraisal of regulation. Ultimately, as signalled by Waller, PVP was adopted without much experience of its effects on the level of drug iatrogeny. The arguments about public health and regulatory failure underpinning PVP again show that the invisible college paid as much attention to a system of control and state of a domain as to solving a particular class of adverse event. It also shows that the college could relate to policy objectives and engage with top regulators' and politicians' agendas for pharmaceutical risk management.

In sum, PVP scientists contribute a great deal to the constitution of pharmaceutical regulation as a domain, a bounded space of issues dealt with systematically by the same people through the same tools. If they finally benefited from the introduction of PVP, it was not only because it protected their own specific jurisdiction of post-marketing safety or pharmacovigilance, but also because it forced them to change their expertise and this jurisdiction into one of 'risk reduction' and 'risk management'. Thus PVP benefited these professionals precisely because they aimed to promote themselves as specialists of medicine safety *in general*, by taking responsibility for the whole life cycle of medicinal products and the outcome of the regulatory domain. Drug safety experts accepted, even pushed in some way for, the redefinition of pharmacovigilance knowledge and practices to become one of risk management. The risk-management programme created a new space of intervention – planning or preparation – by means of which they gained some territory over pre-marketing safety specialists. Absorbing statistical thinking to design clinical trial methodologies, epidemiology to found pharmacoepidemiology, and now audit and information system management techniques to devise industrial pharma-covigilance are all examples of the same phenomenon; that is, the reinvention of medical expertise through hybridisation.[3] PVP experts have gained by not defending the specificity of their expertise in pharmacovigilance but instead transforming it into an overall expertise in risk management.

The circulation of these drug safety evaluators allows for coverage of the control system through practical experience and conceptual constructions. The ability to implement far-reaching changes in the regulation regime is linked to the responsibilities that the scientists of the invisible college face as professionals of medicine safety. PVP reflects a situation in which a profession – medicine – and drug safety scientists can claim large responsibility for the control of medicine safety. Medical expertise covers all medicinal products since medicine as a profession is about therapy and no therapeutic tool escapes its responsibility and scrutiny. Evaluators assess products and are involved 'downstream' in the products' life cycle; that is, at the level of their prescription and use. Being involved in post-marketing surveillance, their knowledge also covers forgotten health risks and the real risks of regulatory failure. It also includes the major modes or tools of control since medical professionals cover the whole life cycle of medicines, being involved in their development and testing, evaluation, use, and surveillance of their use and of side-effects. Doctors have displaced other expertise by inventing new forms of control. Toxicologists have been relegated by the emergence of clinical trials while pharmacists have come second to doctors in monitoring and controlling adverse drug reactions. The invisible college has mostly comprised those with hands-on experience in pharmacovigilance processes. The scientists who standardised PVP had the advantage of being professionally involved on a daily basis with the serial treatment of adverse reactions and a responsibility for controlling them. Post-marketing drug safety evaluators cognitively assemble all of the elements that are part of the control of drug safety. The cognitive span of their college seems greatest of all three cases as the circulation of these scientists across the different spheres of control brought about the sharing of common experience to be formalised in standardised concepts.

BOUNDARIES OF RESPONSIBILITY AND OF STANDARDISATION

The two other cases of HACCP and PMM may be grouped here. Both illustrate, in contrast with PVP and pharmaceutical regulation, that invisible colleges of standardisation are less consequential in situations where they fail to work with all actors of the domain.

The HACCP concept has moved a long way from its initial site of experimentation and conceptualisation, the contract between NASA and the food company Pillsbury, to become an international standard applying beyond confined spaces like submarines and space shuttles. HACCP is applied across the board to a variegated class of products and food operators, ranging from primary producers, farmers, and caterers to canteens. It prescribes the procedures for defining risks and safety issues, and its scope is so large as to be indefinite.[4] This means that HACCP supplements or even generates as shown in Chapter 5, the production of parallel norms adapted to particular parts of the food chain or types of production. National governments, trade associations and more importantly, food distributors have their own guidance for application of the seven HACCP principles. One unique standard in effect is applied through a diversity of sub-regimes. HACCP in this sense does not promote the integration of a food regulation as a domain.

It does not do so either because it made the relation between food businesses and their controllers more complex, or failed to attend to subsequent developments of this relation. Theoretically, as mentioned earlier, the plan was indeed to liberalise food hygiene by alleviating the interventions of official inspectors and increase self-regulation in the area. But HACCP did not simply allow the delegation of food control to businesses or redefine the work of inspectors from one of testing products to the verification of HACCP plans. The new 'advisory' and supervisory role of inspectors appears more complex in practice (Smith-De Waale, 1996), often resulting in local negotiations between them and the manufacturer around the choice of critical control points. Many guidelines have been elaborated in the name of the seven HACCP principles, forming a web of constraints that food businesses can hardly hope to escape and turning an initially quite liberal approach into a prescriptive one (Antle, 1998). In practice, HACCP has opened a large market for the certification of food businesses (Tanner, 2000), and the training of food operators and food hygiene consultants to sell packaged HACCP models or prepare businesses for certification, a result that the ICMSF could only note and regret. All of these intermediary controllers have their own way of adjusting HACCP to the context in which they operate and have often defined particular norms. Substantial standards in the form of good manufacturing practices, critical control points or contamination limits have also been established by regulatory or industry bodies alongside HACCP principles. These developments explain why there are controversies among HACCP specialists as well as food hygiene practitioners on the best combination of HACCP principles with specific product standards (Hathaway, 1995).

Given the ambiguities around its effect, it seems inappropriate to say that a coherent regulation regime has come about with the HACCP concept. There is little indication that HACCP proponents aimed for this or had the capacity to do so. Food microbiologists used a relatively narrow parcel of food regulation: food hygiene in large production line businesses and cases in which contamination is relatively easily to trace or attribute to one source. HACCP's history shows that food microbiologists who modelled this tool drew their experience from such large companies: organisations that are usually highly rationalised with production line and quality control plans in place for producing a variety of foods. Many microbiologists understand the frustration of establishing and applying particular microbiological criteria of contamination for each type of food. They are only too aware of the costs of enforcing detailed prescriptions of food hygiene for veterinary inspectors and food producers. They appreciate the capacity of the principles and terms elaborated by NASA and Pillsbury as a new generic form of food hygiene that travels across sectors of food production and geographical spaces. As consultants, they tested and trialled the HACCP principles before publicising their virtue. It is this relation to a then limited practice of food assurance systems that has given food microbiologists an advantage in constructing this generic practice to reduce food contaminations. But that does not adequately apply to other parts of the food chain, as recognised quite late in the day (Sperber, 2005).

The process of modelling seems a typical strategy of food microbiologists to gain influence in the international organisations that they seek to advise. Quite tellingly, many members of the invisible college of HACCP such as Jean-Claude and Michael, turned to risk analysis after HACCP was adopted (Buchanan, 1995; Walls, 1997; Brown et al., 2002), reflecting the bias towards systems thinking and the tendency to model regulatory intervention on the basis of limited practical experience. A key discussion concerning HACCP was the possibility that substantial standards were set at the national level to define for a particular product which points of the production line could be elected as critical control points, and what thresholds of contamination to set – the so-called 'system-wide risk assessment' (NRC, 1985b; Hathaway, 1995). Systems thinking and modelling were extended to design a policy by which local business food safety systems, in a HACCP format, could be linked to nationwide processes of assessment of risks. This would have a two-way correspondence: data emanating from monitoring activities in businesses would be collated to inform risk assessments, and the values set through quantitative microbiological risk assessment would be used by businesses to establish more scientific HACCP plans. A representation of this grand scheme of the government of food safety can be found in Notermans et al. (1996) in the form shown in Figure 8.1.[5]

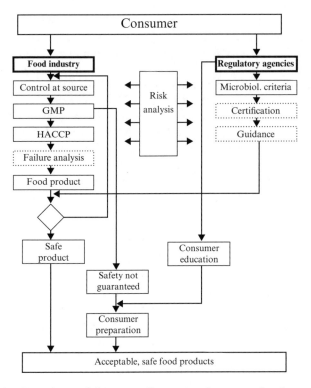

Figure 8.1   Paradigm of the generally accepted concept for the production of microbiologically safe food (reproduced from Notermans et al., 1996: 177)

The concept of a food safety system, after that of HACCP, shows that policy modelling is indeed an approach that food microbiologists favour to reflect on which rules to adopt to reach generally accepted policy objectives (such as Figure 8.1's 'acceptable, safe food products'). It is also by definition geared towards producing highly abstract and conceptual policy schemes. The disadvantage is that it diffracts into a variety of complementary local regimes and produces more unintended and unanticipated effects.

The invisible college did not evaluate the effects of HACCP on the regulation of food safety. HACCP experts are not epidemiologists so they do not spend their time calculating integrated levels of risk, prevalence and incidence of food contaminations – even if individually they are keen to boast about their experience of contaminations. But the latter experience mainly involves contaminations with branded or traceable products and clear causal chains, contexts in which end-of-chain testing could be easily shown to be faulty. There are however many other causes of food contaminations not linked to food production lines and unexplained by the limits of end-of-chain testing. The evaluative attitude of HACCP experts therefore did not extend to all possible adverse events in food consumption, much less to the chaotic effects of the general introduction of HACCP. Even though HACCP is associated with international harmonisation of food standards and food trade liberalisation, its experts did not really try to evaluate the effects of HACCP until it was fully adopted. Other researchers now do the job (Eyck et al., 2006; Cormier et al., 2007; Wright and Teplitski, 2009).

HACCP's position has been to defend the model rather than search for its limits. It was not designed to be an integrated system of control and its experts did not become the central actors on food safety. Food microbiologists have not gained much from the standardisation of food hygiene and its management relative to other disciplines and professional expertise. They have of course partly gained from the introduction of HACCP and the opening of a new intervention in which competences in microbiological analysis have been allied with quality assurance and managerial approaches. Much like PVP, the increasing abstraction of food hygiene techniques using HACCP has given a great deal of value to the expertise of food microbiologists and veterinarians who were trained in these approaches. These disciplines have long pretended they owned the subject of HACCP in the first place because it was developed to deal with biological contaminations of food and was historically pushed by food microbiologists and the ICMSF. The transformation of HACCP into a higher order model of food safety marks attempts by microbiologists to demonstrate how their own expertise could serve to regulate food hygiene. These include the harmonisation of hygiene practices across a highly differentiated food industry, determination of responsibilities in cases of food contamination, and determination of harmonised microbiological criteria in the context of increasing international food trade. However food microbiologists have no precedence in this area that is increasingly recast as food risk management. Even if their expertise is required to resolve this market for business services, others such as toxicologists and quality assurance engineers are equally present. More importantly, food microbiologists did not become, if they ever intended to, the central experts on food safety. The issues they typically act on continue to be seen as just another of the many possible targets that a food risk regulation regime should have.

Again, the restricted circulation of food microbiologists across the different areas of food control explains these limits in food regulation. In contrast to drug safety evaluators, food safety scientists do not have full responsibility for the monitoring and imputation that form the basis for evaluating rules of intervention. While food microbiologists and veterinarians have taken a large role in food hygiene and food control, it is hardly comparable with PVP. They provide the staff of official food inspectorates in many European countries, developing the necessary expertise to design sampling plans and methods of analysis. They are also hired by the food industry for work that complies with food hygiene standards. They populate commissions and committees that set the substantial and performance standards applicable to food businesses from microbiological criteria to food hygiene prescriptions. They have also acquired a decisive role in animal health and control of zoonoses and subsequently in animal diseases transmitted via foods. Historically however, food microbiologists have never been able to cover the whole domain of food control, in breadth or in reach. First, they have never fully embraced issues of chemical contamination of foods. They are in competition with those who work on this other dimension of food safety, such as toxicologists. The second limit to their expertise concerns enforcement and control of food safety beyond areas of primary production. In the past, food businesses operating closer to the final consumer such as distribution and preparation or hospitality have been covered by other inspection services and administrations, in which food microbiology and veterinary knowledge are not as central as legal and economic competences to anticipate and penalise fraud related to price, consumer deception and unfair competition.

Elite food microbiologists tried to reform the domain via a management-based tool of food control, because production line food companies and microbiological issues were their main practical experience, terrain of risk evaluation and source of regulatory influence. But certain food safety issues (chemical contaminations), actors of control (inspectors) as well as sectors of food production and preparation, all components of the fragmented domain of food regulation, were not attended to in their conceptualisation. Food scientists do not assume responsibility for the whole domain as drug safety scientists do and their standardising action does not integrate the domain.

The post-market monitoring of novel foods is an example of limited changes in a system of control, because of a narrow scope and reluctance to prescribe finite protocols to coordinate the applicant company, scientific advisers and decision-makers. From a tool with a large conceptual validity that may apply to all sorts of products, a restrictive standard was formed in which the criteria of intervention were reduced to eight. So even if the standard is defined to apply to novel foods, in practice it singles out a limited series of products, many being unknown at the time of introduction of the standard. It implicitly makes reference to various monitoring experiments, notably that of Unilever which has become a methodological benchmark. On certain aspects, this benchmark is quite precise, including rather fine settings and typified practices for questioning consumers and performing causal analysis on very incomplete data collected through call lines, for instance.

Beyond this however, the standard is confined to risk assessment without considering risk management or the grey area between the interpretation of the results of PMM studies and the ensuing definition of safety specifications and other risk-management measures. The scientists who developed the guidance, 'risk assessors' or scientific advisers to national or European regulatory authorities themselves, argue that PMM is a risk-assessment tool that did not lead automatically to regulatory decisions to authorise or withdraw a product. The latter was paradoxically left in the hands of risk managers. For all the progress it achieved in designing a process of evaluation and monitoring, PMM did respect and refrain from codifying political intervention. There was in this case a well-noted reluctance to set risk-management standards or organise the making of regulatory decisions on the distribution of food products too directly. The limited scope means that there are few such studies. The standard does not justify the establishment of risk-monitoring departments within food companies, or redefine toxicologists in industry or in regulatory agencies as post-marketing safety officers.

The same reluctance to evaluate the action towards policy objectives by food microbiologists is illustrated here, in the restriction on the use of PMM for the verification of safety hypotheses, and its inscription in a risk-analysis framework by means of which the requesting of PMM studies is turned into an uneasy local negotiation between bureaucrats and scientific advisers. Both points show that food safety scientists are reluctant to generalise a standard and install it as an instrument that would indicate that there are large uncertainties and side-effects associated with novel foods, in line with what certain segments of the public think. More strikingly, PMM was not developed against any other tool and is not as such a reform of the regulation of novel foods. Its establishment was the occasion to recall the pre-eminence of the control of novel foods via pre-market assessment. Another new tool to control the efficacy of products came along with PMM, the assessment of nutritional and health claims. In terms of quantified generic risk for novel foods, the main qualitative indicator used by scientists including those active through the invisible college, is the idea that adverse effects of biotechnological foods would have been known if there were any, given the many millions of portions already consumed by American consumers and cattle.

Logically then, the PMM standard did not create any effect on the role of scientific advisers or risk assessors that the members of the ILSI group could have wished for themselves. The scientists of the invisible college had the opportunity to create an overarching instrument for novel foods, put a foot in the door of risk management by setting more extensive criteria of PMM use, and redefine the role of scientific advisers. But they did the opposite in calling PMM a risk-management tool that is part of risk assessment. This reflects a cautious, if not shy, corporatist relation of food safety scientists towards decision-making, and a search for clearer distinctions between their role in regulatory intervention and that of politically accountable actors. In other words, the circuit within which these scientists evolve is smaller and less rich in experience of food risk management. Their opinions should not include the commercial distribution of a product or the definition of a measure or level of protection of consumers, which limits their experience particularly with regard to the commercialisation, distribution and consumption of foods. The relation to foods continues to be an experimental one mediated by tests and protocols, and shaped in academic or industry laboratories. Scientific experts are effectively policed in their behaviour by a boundary between science and politics, and approach decision-making as a no-go area (Roqueplo, 1995; Barthes and Gilbert, 2005; Pielke, 2007; Granjou and Barbier, 2009). These norms denote the need to carve out a space for political decision-making where qualitative judgement on non-chemical or biological parameters can help decide what to protect people from and how to do it. This is a space in which the scheme of logic and formal criteria invented by scientists tend to be rejected. Political decision-making marks the limit of their involvement. The rationalisation of science-based decision-making through principles of risk analysis and the precautionary principle is the sign that the scientific limitations in addressing uncertainty are publicly acknowledged and that policy criteria are superimposed on experts' own criteria (Fisher, 2007; Everson and Vos, 2009).

EMERGENT STANDARDISATION

The detailed analysis of the three cases shows that there is a homologous correspondence between the responsibilities of scientists to evaluate the effectiveness of a system of control – the extent to which it is likely to fail – and their capacity to circulate among the different spheres of regulation. This is why the invisible college of PVP mattered more than the others in the reorganisation of a system of control. It is part of a domain in which the evaluation of adverse events is inseparable from that of the behaviour and operations of other participants – the industry, medical profession, and regulatory agencies. In this domain, those who evaluate are given more responsibility over the avoidance of overall failures in the control of product benefits and risks. This means that standardisation emerges from regulatory domains. By their conceptual constructions, invisible colleges seem to search for ways of reproducing and maintaining systems of control faced by new issues and challenges – rather than altering these systems of control arbitrarily and externally. Invisible colleges appear to be a form of protection of regulatory domains, a way to preserve their boundaries and internal structures in the context of change.

One element anchors this impression; it is the fact that the concepts eventually adopted reinforce the position of actors dominant in the domain. The reform they promote comes from past experiments, thus from the actions of those who are already involved and established in the domain. For instance, pharmacovigilance and drug safety experts are sometimes accused of protecting products and of being reluctant to take them off the market. This accusation was raised in the aftermath of cerivastatin. It is also voiced by general physicians who, when asked by surveyors based in regulatory agencies to detail more information about the conditions of the patient, the reasons for prescribing the drug, the moment when and where a reaction appeared, and so on feel they are simply asked too much for the sake of preserving the status of a product. To what extent this is generally true of all pharmacovigilance activities in all countries remains to be seen. It only shows that pharmacovigilance has become a system mediating the interests of the industry and public authorities to anticipate and respond to public concerns about drug safety.

Standardisation continues a process of industry benchmarking. Regulatory authorities have generalised certain practices innovated in particular parts of the market. Waller's critique of the rush to invent and adopt new tools at the expense of fuller experimentation of their public health benefits could well be the sign of a more structural domination of those actors who favour regulatory harmonisation. This would include dominant players such as representatives of the large pharmaceutical and biomedicine companies, or Big Pharma, and the three main regulators.[6] In the HACCP case, the convergence between the elaboration of the seven principles and a deregulatory and trade-liberalisation policy agenda is striking. The case of post-market monitoring expresses a similar convergence between experts' work to produce concepts and the need of multinational companies to set internal benchmarks and avail themselves of official guidelines. Invisible colleges have strengthened the position of another actor: independent regulatory agencies. National and European independent regulatory agencies such as the EMEA and the EFSA were important nodal points for the constitution of invisible colleges and the standardising of regulatory concepts, as the case of the MCA and PVP shows, as much as the supportive relation of the EFSA to the ILSI in the development of a guideline for PMM. Agencies and colleges are tightly coupled. The literature on agencification (Pollitt et al., 2001; Levi-Faur, 2005) shows that the proliferation of agencies cannot be dissociated from the growing transnational interdependence between scientific experts and the shaping of common standards through these webs that define what an agency is and how it functions (Eberlein and Grande, 2005; Demortain, 2009; Hauray and Urfalino, 2009: 433).[7]

While transnational companies and networks of agencies are confirmed as actors in a domain by invisible colleges, politicians stay at the edge of it – perhaps by the same token. By delegating to these agencies and indirectly to transnational circles of expert scientists, politicians effectively give more autonomy to expert scientists and remove their own opportunities to guide standardisation. Conversely, if scientists assume more than their share of interest and responsibility for the workings of systems of control, this is because margins of manoeuvre are granted to them by the reluctance of other actors – top civil servants, leading national regulators and politicians. The accountability and responsiveness of national public authorities for the outcome of regulatory intervention declines as they become more transnational, systemic and disconnected from public demands and values (Marquand, 2004). No government actor routinely assumes moral or legal responsibility for the whole set of regulatory operations and its outcomes. Invisible colleges are synonymous with the removal of regulatory functions from politicians and representatives of the public.[8] It is thus quite unlikely that the transnationalisation of regulation counters the decrease in civic professionalism observed by Fischer (2009). There will also be little challenge to the disenfranchising of the public from regulatory intervention that students of risk analysis and regulatory science observed very early on (Wynne, 1992; Irwin et al., 1997; Shrader-Frechette, 1997).[9]

An obvious conclusion of this book could have been that colleges are invisible because they are illegitimate. Invisibility – or secrecy in this perspective – would be the protection of an essentially technocratic power group from public scrutiny and legitimacy challenges. But the reverse is probably truer: colleges are highly legitimate for other regulatory actors if not for the public, because they are invisible. Invisibility does not result from power strategies, but from the transversal movements of scientists across the regulatory domain and non-affiliation to one particular vested interest. This is instrumental to the whole domain in constructing and remedying regulatory failures. Invisible colleges provide a space in which regulatory responsibilities can be discreetly evaluated and reformed. In a context in which any new adverse event can reactivate responsibilities, accountability and blame, it is in the interest of everyone and primarily of scientists, to keep standardisation enterprises discreet and to let them be conducted by the insiders of a self-created space for reflection and formalisation of responsibilities. It is quite striking given the level of scrutiny on experts that invisible colleges are not held accountable for the setting of these standards and for their effects. Not much is known, let alone said publicly, about colleges and standardisation. Conceptual standards and their proponents overall command little public attention. Rarely are their origins traced. Scientists are not blamed for having contributed to the passing of such standards as PVP, PMM or HACCP. Organisations like the CIOMS, ILSI or ICMSF are rarely named in the press.[10] Invisible colleges are less a continuation of the transfer of policy from politics to expertise than a sign that regulatory intervention is negotiated through discreet circuits. This allows a regulatory domain to be slowly reformed to adapt to changing issues with minimal challenges from the public and the media. The invisible college phenomenon is kept in the shadows by all of those who have an interest in using scientists' wide-ranging experience to resolve regulatory issues.

Invisible colleges act politically because they allow regulation to function, in spite of perhaps greater complexity implied by multiple intervening actors, issues and demands. There is a strong drive nowadays for more transparency as well as inclusion of the public. But there are also strong indications that invisible colleges will remain. The more disasters there are, the more likely it is that interstitial spaces will emerge and that people will circulate, recreating the invisible college phenomena. This is perhaps counter-intuitive. Disasters are conducive to political crises and public accusations of failure and to formalising responsibilities and lines of accountability. But this only happens when a disaster is proven to be the result of the regulation, which in turn is only possible if there are those in responsible positions who can make the evaluation and are interested in buffering the impact of regulatory disasters. This means that invisible colleges are more likely to appear in areas with high ambitions of control and reliability where risks of failure are high. As in other domains like aviation or nuclear energy, the pharmaceutical industry is a technological area that faces constant risks of technological and regulatory failure, and has indeed faced a long string of disasters and major adverse events. Propelled by the medical profession, these have led to the invention of new tools and systems of control of safety despite the industry's temptation to evade regulation. However, they could not be introduced without common agreement of public authorities and industries. Marketing authorisation, spontaneous reporting systems, and PVP now, help both sides to face their responsibilities in symmetrical ways. The intertwining of responsibilities and the correlative need for mediating actors thus seem greater in domains that have a history of global failures.

More transnational activity also means more interstitial spaces from which to work and potentially again more invisible colleges. Transnationalisation offers more opportunities for reflexive, evaluative actions on regulation regimes. It is also synonymous, as shown in Chapter 7, with greater circulation between spheres and the blurring of boundaries, leading to more confused lines of accountability and responsibilities. In many ways, the transnational regulation of risk is paradoxical; a peculiar situation of uninvolved governance by which no one seems to be ready to take the responsibility for the outcomes of the system. No actor routinely assumes the moral or legal responsibility for the whole set of regulatory operations and their outcomes. Of course, politicians may be blamed for what regulation fails to achieve. But even that accountability works for events that have gone particularly wrong and can be unequivocally attributed to the system of control. Furthermore, blame-avoidance is a common behaviour that has its own effective mechanisms.

So, invisible colleges are 'emergents' from the dynamics of a domain with transnationalisation and complexification of lines of responsibility for failures. 'Emergent' does not mean that action is absent but that it is contingent on certain evolutions of the domain as a whole, which invisible colleges can accentuate. Standardisation is not purely the product of a system. Scientists collectively play an active role and take their share of responsibility, even if in producing these new concepts they also have a chance to moderate this responsibility and accountability. Invisible colleges are neither irresponsible carriers of models nor policy entrepreneurs, but collectives formed in political circumstances. They provide an occasion for scientists to opportunistically fulfil the political vocation that the modern logic of specialism frustrates. Much like Boyle managed to accomplish his mixed experimental and utopian penchants simultaneously by engaging with varied and colourful characters who formed his very own invisible college, contemporary invisible colleges are the structures through which scientific experts can informally and somehow invisibly, fulfil an inclination to participate in regulation and governance beyond the boundaries of their professional space. This form of political action is constitutive of science in a pre-modern sense before it fully embraced ideals of rationality and objectivity. It does not redefine who scientists are and why they act. The action is contingent and unstable. It is a weak rather than a strong commitment confined within their limited but nonetheless effective legitimacy as regulatory professionals.

NOTES

1. Even more so than law-makers and judges who use fairly broad criteria for defining a medicine. The criteria have also been expanded by the European judge to include the strict assessment of anything resembling a medicinal product – certain food supplements for instance (Daburon-Garcia, 2001).
2. In Europe, authorising a product is officially a decision of the European Commission, following a vote in a regulatory committee comprising member state representatives. This vote, in practice, consists of approving the decision prepared by the EMA secretariat on the basis of scientific discussions among members of its committees including the Committee for Human Medicinal Products. The boundary between the making of a decision by civil servants and government representatives on the one hand, and scientific assessment by medical experts on the other, is not clearly demarcated. The vote rarely, if ever, goes against the decision recommended by the EMA and scientific experts (Gehring and Krapohl, 2007; Demortain 2008; Hauray and Urfalino, 2009). Manufacturers and national governments direct their attempts at influencing the process at the EMA and around the evaluation of scientific committees rather than at the stage of formal decision-making. Scientific experts go beyond the strict interpretation of the data to explicitly discuss the need for the product under evaluation. These discussions involve general conceptions on the social value of medication (Hauray, 2005). Regulatory decisions are thus based on expert judgement, without this arousing as much contestation as in the evaluation of food or chemical products.
3. See Kurunmäki (2004) and Miller et al. (2008) for another obvious example of the hybridising of medical expertise and accounting in the context of introduction of cost accounting in Finnish and UK hospitals.
4. The HACCP concept has strayed beyond the food industry to diverse areas such as the improvement of epidemiological surveillance networks (Dufour, 1999) or safety in the production of medical devices and pharmaceuticals (Jahnke and Kühn, 2003).
5. Scientists played their part in establishing the meaning of a 'food safety system'. In an ILSI-sponsored paper, it is defined as 'the institutions, policies, laws and guidelines for the evaluation of risk' (Schilter et al., 2003). The concept has a family resemblance to a 'food control system' ('the integration of a mandatory regulatory approach with preventive and educational strategies that protect the whole food chain') promoted by FAO as the model for developing countries (FAO/WHO, 2003: 3).
6. Critics of the ICH process argue that it regularly opts for lowering safety standards through accommodating industry's agenda to decrease regulatory pressure. The ICH is thus unexceptional in the capture of regulatory agencies by the industry (Abraham and Reed, 2002).
7. On a related note, the formation of the EMEA is best understood as the scaling up of the evaluation system in Europe to the standard of the two other dominant drug regulators of this globalised industry, the FDA and the Japanese Ministry for Social Affairs. Fernand Sauer (see Chapter 3) – to many the single hero of this success story of regulatory Europeanisation – often emphasises that the gathering of national expert resources was only aimed at ensuring Europe had the means to evaluate innovative and internationally distributed products to join the global dance. The interdependence among the three regional regulators only increased after the creation of the EMEA through coordinated product evaluations and the ICH process. This gives teeth to the analysis of Saskia Sassen (2007) for whom regulatory agencies are not 'national' or even 'European' institutions, but transnational ones; an adaptation of the state machinery – its 'denationalisation' (Sassen, 1996) – to the scale on which markets and products are made. Agencies form a network with their counterparts in other countries to execute transnationalised functions of market regulation (Sassen, 2007).
8. The democratic record of independent agencies is quite poor (Borrás et al., 2007). The EMA confirms the pro-industry bias of EU regulation (Permanand and Mossialos, 2005; Permanand, 2006). Efforts to increase the transparency and responsiveness of

the agency to the public, for instance by opening their board, do not seem overtly effective (Groenleer, 2009). The domain continues to be marked by secrecy and blatant conflicts of interests (Abraham and Lewis, 2000; Abraham and Davis, 2006; Abraham, 2007; Vitry et al., 2008). On the food side, the EFSA is often criticised for dealing with biotechnologies on the basis of an industry-friendly science and in a way that is unlikely to improve the science–society relationship (Borrás, 2006). Together with the European Commission and experts, it forms an institutional coalition that upholds a technocratic approach (Chalmers, 2003 and 2004). Similarly, negative assessments are made of national pharmaceutical and food agencies (Benamouzig and Besançon, 2005; McGoey, 2007; Rothstein, 2007).

9. There is much to be done in the way of reconciling expertise and public participation in transnational or global governance in the way critical sociological and political literature has attempted in national settings (Turner, 2001; Collins and Evans, 2002; Callon et al., 2009).

10. Environmental NGOs, patients' associations or those scientists with a more vocal public health or anti-industry orientation sometimes criticise the ILSI for instance. But such criticism is sporadic rather than systematic and concerns industry's structural influence rather than the content of a particular concept. Ultimately, none of these criticisms have prevented the three concepts studied here from entering into practice and law.

Appendix 1: Research strategy and methodology

The ambition of the research project that resulted in this book was to find the cause(s) of an outcome: the production of standards of intervention by sets of polyvalent and influential scientists. Following Ragin (1989), an appropriate strategy to help the discovery of these causes is to compare cases that are dissimilar in many aspects except for this particular outcome.

The regulation regimes for food and for pharmaceuticals are such a case, as are risk-regulation regimes in general that differ considerably (Hood et al., 2001). The regime of pharmaceutical regulation is typically considered to be quite distinct, featuring a particularly strong influence of a concentrated and internationalised industry, as well as the centrality of medical professionals as intervening actors in a domain that seeks to balance the risks of products with their (therapeutic) benefits. The domain has also long done without any concept of risk, drawing on a more autonomous medical expertise. The concept of risk is now being taken up in the regime, as illustrated by the recasting of safety and efficacy evaluation as 'risk/benefit assessment', of post-marketing surveillance as 'risk management' and the recognition of the importance of improved communication to patients and doctors under the rubric of 'risk communication'. These traits contrast strongly with food regulation, which is often presented as fragmented (per types of foods and modes of control applying to each), less internationalised (regimes remain very much national, in line with the cultural component of food consumption and diets) and also much more politicised (Smith and Phillips, 2000).

These differences supposedly reflect market structures and product conceptions. Medicines are highly standardised products and their composition and purity are strictly controlled. They are mainly composed of one pure substance or molecule, the discovery and development of which costs several hundred millions dollars (DiMasi et al., 2003; DiMasi and Grabowski, 2007), resulting in a global investment/sale ratio of 16.1% (JRC, 2008). The pharmaceutical industry is also dominated by large research-based companies, even if smaller pharmaceutical research companies play a larger role now in the development of innovative products. It

211

is estimated that the 30 largest companies account for around 70% of the 600,000 billion euros worth of worldwide pharmaceutical sales (Clinton and Mozeson, 2009), with sales of half of their products being made on foreign markets. Foodstuffs are highly diverse. Some are standardised in their composition and industrially produced; others are picked or grown. In one class of foodstuff, the composition and nutritional value vary greatly.[1] Furthermore, their production and consumption are not as closely controlled as medicines, and the risks they pose for our health may come from the natural development of bacteria and micro-organisms as much as from production processes. The food industry consists of myriad businesses (310,000 small and medium enterprises in the EU alone), and the world's 50 world biggest manufacturers (in terms of turnover) account for less than 20% of global food production (USDA, 2010). It dedicates just about 1% of its turnover to R&D (JRC, 2008).

Given these differences, it is noteworthy that a growing range of products or substances have to undergo the same sort of rationalised and codified risk assessment, be they (eventually labelled and distributed as) foods or medicines. In short, the domains are different but share a similar outcome in the emergence of a logic of evaluation and more generally, in their scientisation and rationalisation. The regulatory demands for establishing scientifically the benefits or risks of a product do in fact create a convergence between domains and markets. For instance, the demand for proof of the therapeutic efficacy and innocuity of pharmaceuticals has impacted many more types of products, as case-laws and the medical professional have enlarged the criteria of what a medicine is, to make sure substances intended as food supplements or functional foods undergo the same strict evaluation as substances intended for therapeutic purposes. And indeed, the development of functional foods has blurred the boundary between food and medicines. Recently invented products such as energy drinks and cholesterol-lowering yellow fat spreads, still considered part of food, are quasi-medicines if judged by the standardisation of their composition, the concentration of a substance conceived to produce particular effects, and the intention of those who consume these products. More and more food products are assessed following codified tests and trials or through a risk-assessment methodology that has become increasingly formalised over the years and more professionalised as an exercise. In sum, there is an outcome that is common to different areas, and which makes these areas converge. These are the two reasons to engage in this comparison.

Comparability was enhanced by focusing on three concepts, one for each domain (conventional foods, novel or functional foods, medicines). The three concepts were turned into formal international standards around the same period (between 1995 and now). All three are about surveillance or monitoring of risks. They all in their own way represent a shift towards an epidemiological logic of imputation, focused on constructing as realistic an image as possible of the adverse reactions, contaminations and diseases occurring out there, before discerning their causes and attributing responsibilities. They mainly impact large businesses, as the organisation with primary responsibility for this surveillance. But they also orchestrate the relations between a variety of actors who participate in regulatory intervention on products and adverse events, from the different industry units involved (product development, quality assurance, safety officers, etc.) to inspectors, regulatory agencies and the scientific advisers who work for these companies, so bypassing professionals and even consumers or patients as sensors of the risks and transmitters of information and data.

The main approach for studying these concepts is genealogical: it consists of detecting the various places in which they simultaneously emerged, sometimes under slightly different guises, and tracing the connections that were made between these different places to channel the different versions of the concept into a common and standardised one. Three sources were used.

The starting point for the research is to look at the standards as defined in European legislation and official guidelines, as well as in the many different types of less legal but nonetheless official texts that make reference to the concepts (industry guidance, scientific opinions published by scientific committees, policy papers from regulatory agencies.[2]

Second, scientific publications were used. Scientific publications are helpful to give a full account of the dispersion of scientific research on the standards. The main methodological innovation for this research was in analysing the journal articles published on the subject, which defined the standards, constructed the adverse events and regulatory theories that formed its content, referred to the places, organisations and people that pushed or resisted them, and so on. Journals of toxicology, microbiology, pharmacology and epidemiology regularly publish articles on PVP, PMM and HACCP. These articles are landmarks in the running history of disciplines and were used as a source of basic information about the history of these standards. This information helped to analyse the trajectory of regulatory concepts, from the day and place where the practices were first named to the inscription of the concept in the text of a formal standard. Scientific publications led me to identify important points in time and the sites in which the concepts were critically discussed, renamed and reinforced, and to know who the scientists that were present were.

Two semi-quantitative methods were used on this information. In

Chapter 3, I present a co-word study of the field of evaluation research. The co-word methodology consists of measuring the number of times two words appear together in a corpus or co-occurrence. This method is based on the linguistic notion of semantic fields. It helps to distinguish the areas in which they are more dense as well as the boundaries. The co-word method is an alternative to citation software, as it does not have the disadvantage of leaving certain disciplines or sub-disciplines and authors in the shade on the grounds that they are not cited. An important proportion of papers that are referenced in large databases do not appear in maps of co-citation networks. Being strictly quantitative, co-citation research does not facilitate tracing the production of knowledge, its dynamics and conflicts (Callon et al., 1993; Leydesdorff, 1989; Courtial, 1994). In Chapter 6, I use a second type of scientific research mapping, that of co-authorship networks on the basis of the publications relating to each standard in the Medline or Web of Science databases, retrieved by using the name of the concept as a topic keyword. When cross-checked with qualitative information about informal collaborations among scientists that were extracted from interviews, they allow for the characterisation of the types of colleges involved.

Both types of maps were produced by the Réseau-Lu software developed by Andrei Mogoutov in association with Alberto Cambrosio at the University of McGill, and later with researchers of the Centre de Sociologie des Innovations of the Ecole des Mines de Paris. Réseau-Lu treats quantitative information visually through maps that represent the relations between the items in a database as well as the strength of these relations (two elements clustered on the map are strongly related; the analyst can choose to visualise more or less strongly associated items, to focus or unfocus the map). The software reconstructs and visualises this information in the form of networks and has strong undertones of social network analysis. It is a highly effective tool to embrace, visually, large corpuses of data (Cambrosio et al., 2004).

Finally, 18 oral history interviews were conducted with 16 scientists who appeared to have played a central role in the adoption of the standard, in order to reconstruct the history of development of the standard.[3] The scientists who were selected were both authors of relevant academic papers and active as scientific advisers to governmental bodies or regulatory agencies. Questions were also asked more specifically about the careers of these scientists, their relation to activities as 'experts', their collaborations, and so on. Thirty-seven interviews were conducted in parallel in regulatory agencies and in the industry in several countries and at the EU level to learn more about the practices that were being standardised, the impact of the adopted standards, and also their policy context.

NOTES

1. In moderation, it should be noted that food distribution and consumption across countries follow increasingly similar patterns (see for instance Regmi et al., 2008).
2. These standards are also cited in innumerable Internet pages – in such quantity that this source has been put aside, except for general information and exploration purposes.
3. The information collected through these interviews is used throughout the book and is indeed the main material for the study of the three cases. Particular interviews are mentioned only when I quote the interviewee.

References

Abbott, A. (1988), *The System of Professions. An Essay on the Division of Expert Labour*, Chicago: Chicago University Press.

Abraham, J. (2007), 'Democracy, technocracy and pharmaceutical regulation', Paper presented at the conference on Health Security Agencies Between Technocracy and Democracy, University of Liège, Belgium.

Abraham, J. and Davis, C. (2005), 'A comparative analysis of drug safety withdrawals in the UK and the US (1971–1992): Implications for current regulatory thinking and policy', *Social Science & Medicine*, **61**(5), 881–892.

Abraham, J. and Davis, C. (2006), 'Testing times: The emergence of the Practolol disaster and its challenge to British drug regulation in the modern period, *Social History of Medicine*, **19**(1), 127.

Abraham, J. and Lewis, G. (2000), *Regulating Medicines in Europe: Competition, Expertise, and Public Health*, London: Routledge.

Abraham, J. and Reed, T. (2002), 'Progress, innovation and regulatory science in drug development: The politics of international standard-setting', *Social Studies of Science*, **32**(3), 337–369.

Adams, C. E. (2002), 'Hazard analysis and critical control point – original "spin"', *Food Control*, **13**(6), 355–358.

Adler, E. and Haas, P. M. (1992), 'Epistemic communities, world-order, and the creation of a reflective research-program – conclusion', *International Organization*, **46**(1), 367–390.

AFSSAPS (2001), *Rapport sur les conditions de retrait du marché des spécialités contenant de la cérivastatine (Staltor® & Cholstat®) le 08 août 2001 et obligations de pharmacovigilance*, Saint-Denis: AFSSAPS.

AFSSAPS (2009), *Risk management activity at AFSSAPS: Organisation, functioning, partnerships and developments*, Saint-Denis: AFSSAPS.

Alemanno, A. (2007), *Trade in Food: Regulatory and Judicial Approaches in the EC and the WTO*, London: Cameron May.

Amsterdamska, O. (2009), 'Microbiology', in P. J. Bowler and J. V. Pickstone (eds), *The Cambridge History of Science: The Modern Biological and Earth Sciences* (Vol. 6), Cambridge: Cambridge University Press, pp. 316–341.

Andrews, E. and Dombeck, M. (2004), 'The role of scientific evidence of risks and benefits in determining risk management policies for medications', *Pharmacoepidemiology and Drug Safety*, **13**(9), 599–608.

Andrews, E. B. (1997), 'The re-emergence of epidemiology within the pharmaceutical industry', *Pharmacoepidemiology and Drug Safety*, **6**(1), 57–59.

Ansell, C. and Vogel, D. (2006), 'The contested governance of European food safety regulation', in C. Ansell and D. Vogel (eds), *What's the Beef? The Contested Governance of European Food Safety*, Cambridge, MA: MIT Press, pp. 3–32.

Antle, J. M. (1998), 'The cost of quality in the meat industry: Implications for HACCP regulation', in *Proceedings of the NE-165 Conference on Economics of HACCP*, 15–16 June 1998, Washington, DC.

Appel, W. C. (1998), 'Harmonization in regulatory pharmacovigilance: Impracticalities and scientific irrationality', *Pharmacoepidemiology and Drug Safety*, **7**(5), 359–361.

Arai, S. (1996), 'Studies on functional foods in Japan: State of the art', *Bioscience, Biotechnology, and Biochemistry*, **60**(1), 9–15.

Arlett, P (2004), *Pharmacovigilance: Common Provisions, Private Communication*, Brussels: TOPRA.

Arlett, P. and Harrison, P. (2001), 'Compliance in European pharmacovigilance: A regulatory view', *Pharmacoepidemiology and Drug Safety*, **10**(4), 301–302.

Arlett, P., Moseley, J. and Seligman, P. J. (2005), 'A view from regulatory agencies', in B. L. Strom (ed.), *Pharmacoepidemiology*, 4th edn, Chichester: John Wiley & Sons, pp. 116–117.

Ashford, N. A. (1984), 'Advisory committees in OSHA and EPA: Their use in regulatory decisionmaking', *Science, Technology, and Human Values*, **9**(1), 72–82.

Ashwell, M. (2003), *ILSI Europe Concise Monograph on Concepts of Functional Foods*, Brussels: ILSI Europe.

Atkin, L., Bauman, H., Jezeski, J. and Silliker, J. (1972), 'Prevention of contamination of commercially processed foods', Paper presented at the National Conference on Food Protection, Washington, DC, US Government Printing Office.

Atkinson, M. M. and Coleman, W. D. (1989), 'Strong states and weak states: Sectoral policy networks in advanced capitalist economies', *British Journal of Political Science*, **19**(1), 47–67.

Baird-Parker, A. C. (1995), 'Development of industrial procedures to ensure the microbiological safety of food', *Food Control*, **6**(1), 29–36.

Baldwin, R. and Cave, M. (1999), *Understanding Regulation: Theory, Strategy, and Practice*, Oxford: Oxford University Press.

Baldwin, R., Scott, C. and Hood, C. (1998), *A Reader on Regulation*, Oxford: Oxford University Press.

Bardach, E. (1989), 'Social regulation as a generic policy instrument', in L. M. Salomon and M. S. Lund (eds), *Beyond Privatization: The Tools of Government Action*, Washington, DC: The Urban Institute, pp. 197–230.

Barnes, J. and Mitchell, R. T. (2000), 'HACCP in the United Kingdom', *Food Control*, **11**(5), 383–386.

Barthes, Y. and Gilbert, C. (2005), 'Impuretés et compromis de l'expertise. Une difficile reconnaissance', in L. Dumoulin, S. La Branche, R. Cecile and P. Warin (eds), *Le recours aux experts. Raisons et usages politiques*, Grenoble: Presses Universitaires de Grenoble, pp. 43–62.

Bauman, H. E. (1974), 'The HACCP concept and microbiological hazard categories', *Food Technology*, **28**(9), 30–32.

Bauman, H. E. (1986), 'The hazard analysis critical control point concept', in C. W. Felix (ed.), *Food Protection Technology*, Chelsea, VT: Lewis Publishers, pp. 175–179.

Beck, B. (1989), 'The use of toxicology in the regulatory process', in A. W. Hayes (ed.), *Principles and Methods of Toxicology*, 2nd edn, New York: Raven Press, pp. 23–76.

Beck, U. (1992), *Risk Society: Towards a New Modernity*, London: Sage Publications.

Beck, U., Giddens, A. and Lash, S. (eds) (1994), *Reflexive Modernization: Politics, Tradition and Aesthetics in the Modern Social Order*, Stanford, CA: Stanford University Press.

Bégaud, B. (1993), 'Postface – Special Issue: Methodological approaches in pharmacovigilance', *Post Marketing Surveillance*, **7**(1–2), 230–235.

Bégaud, B. (1999), 'L'apport de la pharmacoépidémiologie', *Actualités et Dossier de Santé Publique*, **27**, 33–37.

Bégaud, B., Chaslerie, A. and Haramburu, F. (1994), 'Organisation et résultats de la pharmacovigilance en France', *Epidémiologie et Santé Publique*, **42**, 416–423.

Béjot, J. (2002), *Microbiologie*, Paris: Encyclopedia Universalis.

Bell, R. G. (1993), 'Development of the principles and practices of meat hygiene: A microbiologist's perspective', *Food Control*, **4**(3), 134–140.

Benamouzig, D. and Besançon, J. (2005), 'Administrer un monde incertain: Les nouvelles bureaucraties techniques. Le cas des agences sanitaires en France', *Sociologie du Travail*, **47**(3), 301–322.

Ben-David, J. (1984), *The Scientist's Role in Society,* Chicago: Chicago University Press.

Benford, D. (2000), *The Acceptable Daily Intake, a Tool for Ensuring Food Safety*, Washington, DC: ILSI Press.

Bennett, C. J. and Howlett, M. (1992), 'The lessons of learning: Reconciling theories of policy learning and policy change', *Policy Sciences*, **25**(3), 275–294.

Benveniste, G. (1972), *The Politics of Expertise*, Berkeley, CA: The Glendessary Press.

Berg, M. (1995), 'Turning a practice into a science: Reconceptualizing postwar medical practice', *Social Studies of Science*, **25**(3), 437–476.

Bernard, D. (1998), 'Developing and implementing HACCP in the USA', *Food Control*, **9**(2–3), 91–95.

Berneker, G. C. (1992), 'Workshop on ADR monitoring and assessment: Introduction and background', *Pharmacoepidemiology and Drug Safety*, **1**(5), 213–214.

Bernstein, P. L. (1995), *Against the Gods: The Remarkable Story of Risk*, New York: John Wiley & Sons.

Bigwood, E. J. (1964), *Du problème de l'harmonisation des legislations nationales en matiere de reglementation d'hygiene des aliments. Le droit de l'alimentation dans l'Europe de demain/Food Law in the Europe of Tomorrow*, Brussels: Université Libre de Bruxelles.

Bigwood, E. J. and Gérard, A. (1967), *Objectifs et principes fondamentaux d'un droit comparé de l'alimentation*, Basle: S. Karger.

Black, J. (2002), 'Critical reflections on regulation', *Australian Journal of Legal Philosophy*, **27**, 1–36.

Black, J. (2008), 'Constructing and contesting legitimacy and accountability in polycentric regulatory regimes', *Regulation & Governance*, **2**(2), 137–164.

Blanchfield, J. R. (1992), 'Due diligence – defence or system?', *Food Control*, **3**(2), 80–83.

Bodewitz, H., Buurma, H. and de Vries, G. H. (1987), 'Regulatory science and the social management of trust in medicine', in W. Bijker, T. P. Hughes and T. Pinch (eds), *The Social Construction of Technological Systems: New Directions in the Sociology and History of Technology*, Cambridge, MA: MIT Press, pp. 243–259.

Boli, J. and Thomas, G. M. (1997), 'World culture in the world polity: A century of international non-governmental organization', *American Sociological Review*, **62**(2), 171–190.

Bonnaud, L. and Coppalle, J. (2009), 'Les inspecteurs vétérinaires face aux normes privées', *Review of Agricultural and Environmental Studies*, **90**(4), 399–422.

Borden, E. K. (1981), 'Post-marketing surveillance: Drug epidemiology', *The Journal of International Medical Research*, **9**(6), 401–407.

Borrás, S. (2006), 'Legitimate governance of risk at the EU level? The case of genetically modified organisms', *Technological Forecasting and Social Change*, **73**(1), 61–75.

Borrás, S., Koutalakis, C. and Wendler, F. (2007), 'European agencies and input legitimacy: EFSA, EMEA and EPO in the post-delegation phase', *Journal of European Integration*, **29**(5), 583–600.

Borraz, O. (2007a), 'Risk and public problems', *Journal of Risk Research*, **10**(7), 941–957.

Borraz, O. (2007b), 'Governing standards: The rise of standardization processes in France and in the EU', *Governance*, **20**(1), 57–84.

Börzel, T. A. (1998), 'Organizing Babylon – on the different conceptions of policy networks', *Public Administration*, **76**(2), 253–273.

Botzem, S. and Hofmann, J. (2010), 'Transnational governance spirals: The transformation of rule-making authority in Internet regulation and corporate financial reporting', *Critical Policy Studies*, **4**(1), 18–37.

Boudia, S. (2007), 'Global regulation: Controlling and accepting radioactivity risks', *History and Technology*, **23**(4), 389–406.

Boudia, S. (2010), 'Managing scientific and political uncertainty. Risk assessment in an historical perspective', Paper given at the conference on Carcinogens, Mutagens, Reproductive Toxicants: The Politics of Limit Values and Low Doses in the Twentieth and Twenty-first Centuries, Strasbourg: IRIST, University of Strasbourg.

Bovens, M. A. P. and 'tHart, P. (1996), *Understanding Policy Fiascoes*, New Brunswick, NJ: Transaction Publishing.

Bradbury, J. A. (1989), 'The policy implications of differing concepts of risk', *Science, Technology & Human Values*, **14**(4), 380–399.

Braithwaite, J. (1993), 'Transnational regulation of the pharmaceutical industry', *The Annals of the American Academy of Political and Social Science*, **525**(1), 12–30.

Braithwaite, J. and Drahos, P. (2000), *Global Business Regulation*, Cambridge: Cambridge University Press.

Breyer, S. (1993), *Breaking the Vicious Circle: Toward Effective Risk Regulation*, Cambridge, MA: Harvard University Press.

Brickman, R., Jasanoff, S. and Ilgen, T. (1985), *Controlling Chemicals: The Politics of Regulation in Europe and the United States*, Ithaca, NY: Cornell University Press.

Brint, S. G. (1994), *In an Age of Experts: The Changing Role of Professionals in Politics and Public Life*, Princeton, NJ: Princeton University Press.

Brown, M., and Stringer, M. (2002), *Microbiological Risk Assessment in Food Processing*, Cambridge: Woodhead Publishing & CRC Press.

Brunsson, N. and Jacobsson, B. (2000), 'The contemporary expansion of standardization', in N. Brunsson and B. Jacobsson (eds), *A World of Standards*, Oxford: Oxford University Press, pp. 1–17.

Buchanan, R. L. (1995), 'The role of microbiological criteria and risk assessment in HACCP', *Food Microbiology*, **12**, 421–424.

Buonanno, L. (2006), 'The creation of the European Food Safety Authority', in C. K. Ansell and D. Vogel (eds), *What's the Beef? The Contested Governance of European Food Safety*, Cambridge, MA: MIT Press, pp. 259–278.

Burgess, A. (2006), 'The making of the risk-centred society and the limits of social risk research', *Health, Risk & Society*, **8**(4), 329-342.

Burt, R. S. (1995), *Structural Holes: The Social Structure of Competition*, Cambridge, MA: Harvard University Press.

Busch, P. O. and Jorgens, H. J. (2005), 'The international sources of policy convergence: Explaining the spread of environmental policy innovations', *Journal of European Public Policy*, **12**(5), 860–884.

Büthe, T. (2009), 'The politics of food safety in the age of global trade: The Codex Alimentarius Commission in the SPS-Agreement of the WTO', in C. Coglianese, A. Finkel and D. Zaring (eds), *Import Safety: Regulatory Governance in the Global Economy*, Philadelphia: University of Pennsylvania Press, pp. 88–109.

Buton, F. (2006), 'De l'expertise scientifique à l'intelligence épidémiologique: L'activité de veille sanitaire', *Genèses*, **65**, 71–91.

Caduff, L. and Bernauer, T. (2006), 'Managing risk and regulation in European food safety governance', *Review of Policy Research*, **23**(1), 153–168.

Calderon, R. L. (2000), 'Measuring risks in humans: The promise and practice of epidemiology', *Food and Chemical Toxicology*, **38**(Suppl. 1), S59–S63.

Callon, M., Barthe, Y. and Lascoumes, P. (2009), *Acting in an Uncertain World: An Essay on Technical Democracy*, Cambridge, MA: The MIT Press.

Callon, M., Courtial, J. P. and Penan, H. (1993), *La scientométrie*, Paris: Presses Universitaires de France.

Callon, M., Lascoumes, P. and Barthes, Y. (2009), *Acting in an Uncertain World: An Essay on Technical Democracy*, Cambridge, MA: MIT Press.

Cambrosio, A. and Keating, P. (1983), 'The disciplinary stake: The case of chronobiology', *Social Studies of Science*, **13**(3), 323–353.

Cambrosio, A., Keating, P. and Mogoutov, A. (2004), 'Mapping collaborative work and innovation in biomedicine: A computer-assisted analysis of antibody reagent workshops', *Social Studies of Science*, **34**(3), 325–364.

Canu, R. and Cochoy, F. (2004), 'La loi de 1905 sur la répression des fraudes: Un levier décisif pour l'engagement politique des questions de consommation?', *Sciences de la Société*, **62**, 69–92.

Carpenter, D. P. (2010), *Reputation and Power: Organizational Image and Pharmaceutical Regulation at the FDA*, New York: Princeton University Press.

Castle, W. M. (1992), 'The report of the CIOMS II Working Group', *Pharmacoepidemiology and Drug Safety*, **1**(1), 53–54.

Ceccoli, S. J. (2003), 'Policy punctuations and regulatory drug review', *Journal of Policy History*, **15**(2), 157–191.

Chalmers, D. (2003), '"Food for thought": Reconciling European risks and traditional ways of life', *Modern Law Review*, **66**(4), 532–562.

Chalmers, D. (2004), 'Risk, anxiety and the European mediation of the politics of life: The European Food Safety Authority and the government of biotechnology', *European Law Review*, **30**(5), 649–674.

Champenois, C. (2001), 'Le système allemand d'évaluation des médicaments', Mémoire de DEA de sociologie de l'Institut d'études politiques, Paris.

Chappel, C. I. (1984), 'Address to Professor René Truhaut', in *Cent cinquantième anniversaire de la création de la Chaire de Toxicologie de la Faculté de Pharmacie de Paris et Jubilé scientifique du Professeur René Truhaut*, Paris: Comité d'organisation du jubilé scientifique du Prof. R. Truhaut, p. 221.

Chast, F. (2002), *Histoire contemporaine des médicaments*, Paris: La Découverte.

Chauveau, S. (2004), 'Genèse de la "sécurité sanitaire": Les produits pharmaceutiques en France aux XIXe et XXe siècles', *Revue d'histoire moderne et contemporaine*, **51**(2), 88–117.

Chubin, D. E. (1985), 'Beyond invisible colleges – inspirations and aspirations of post-1972 social studies of science', *Scientometrics*, **7**(3–6), 221–254.

CIOMS (2006), *The Development Safety Update Report (DSUR): Harmonizing the Format and Content for Periodic Safety Reporting During Clinical Trials*, Geneva: CIOMS.

Clark, M. (2000), *Regulation: The Social Control of Business Between Law and Politics*, Basingstoke: Palgrave Macmillan.

Claude, J.-R. (1984), 'Investigations toxicologiques pour la mise sur le marché des médicaments en 1984: Stratégie, nature, validité scientifique', in *Cent cinquantième anniversaire de la création de la Chaire de Toxicologie de la Faculté de Pharmacie de Paris et Jubilé scientifique du Professeur René Truhaut*, Paris: Comité d'organisation du jubilé scientifique du Prof. R. Truhaut, pp. 226–229.

Clergeau, C. (2005), 'European food safety policies. Between a single market and a political crisis', in M. D. Stephens (ed.), *Health Governance in Europe. Issues, Challenges and Theories*, London: Routledge, pp. 113–133.

Clinton, P. and Mozeson, M. (2009), *The Pharm Exec 50*, New York: Pharmaceutical Executive.

Codex Alimentarius Commission (1981), *Report of The Seventeenth Session of The Codex Committee On Food Hygiene, Alinorm 81/13*, Geneva: FAO/WHO.

Codex Alimentarius Commission (1985), *Report of the Twentieth Session of the Codex Committee on Food Hygiene, Alinorm 85/13A*, Geneva: FAO/WHO.

Codex Alimentarius Commission (1993), *Report of the Twenty-Sixth Session of the Codex Committee On Food Hygiene, Alinorm 93/13A*, Geneva: FAO/WHO.

Codex Alimentarius Commission (1996), *Draft Hazard Analysis and Critical Control Point (HACCP) System and Guidelines for its Application, Report of the 29th session of the Codex Committee on Food Hygiene, Alinorm 97/31A*, Geneva: FAO/WHO.

Codex Alimentarius Commission (1997), *Basic Texts On Food Hygiene*, Geneva: FAO/WHO.

Codex Alimentarius Commission (2003), *Recommended International Code of Practice General Principles of Food Hygiene, Cac/Rcp 1-1969, Rev. 4-20031*, Geneva: FAO/WHO.

Coglianese, C. and Lazer, D. (2003), 'Management based regulation: Prescribing private management to achieve public goals', *Law & Society Review*, **37**(4), 691–730.

Coleman, W. D. and Perl, A. (1999), 'Internationalized policy environments and policy network analysis', *Political Studies*, **47**(4), 691–709.

Collins, H. M. (1974), 'The TEA set: Tacit knowledge and scientific networks', *Social Studies of Science*, **4**(2), 165–185.

Collins, H. M. (1981a), 'Introduction: Stages in the empirical programme of relativism', *Social Studies of Science*, **11**(1), 3–10.

Collins, H. M. (1981b), 'The place of the core-set in modern science: Social contingency with methodological propriety in science', *History of Science*, **19**, 6–19.

Collins, H. M. and Evans, R. (2002), 'The third wave of science studies: Studies of expertise and experience', *Social Studies of Science*, **32**(2), 235–296.

Commission Decision 2000/500/EC on authorising the placing on the market of 'yellow fat spreads with added phytosterols' as a novel food or novel food ingredient under Regulation 258/97, *Official Journal L 200*, 8 August 2000, p. 59.

Commission Decision 2001/471/EC of 8 June 2001 laying down rules for the regular checks on the general hygiene carried out by the operators in establishments according to Directive 64/433/EEC on health conditions for the production and marketing of fresh meat and Directive 71/118/ EEC on health problems affecting the production and placing on the market of fresh poultry meat, *Official Journal L 165*, 21 June 2001, pp. 0048–0053.

Commission Decision 93/51/EEC of 15 December 1992 on the microbiological criteria applicable to the production of cooked crustaceans and molluscan shellfish, *Official Journal L 013*, 21 January 1993, pp. 0011–0013.

Commission of the European Communities (1989), Report on pharmacovigilance in the European Community, III/3577/89, September 1989, Brussels.

Commission of the European Communities (1997), The General Principles of Food Law in the European Union, Commission Green Paper, COM(97)176 final, Brussels.

Commission of the European Communities (2002), Opinion of the Scientific Committee on Food on a report on Post Launch Monitoring of 'yellow fat spreads with added phytosterol esters', SCF/CS/NF/DOS/ 21 ADD 2 final, Brussels.

Commission of the European Communities (2008a), Proposal for a Regulation of the European Parliament and of the Council on novel foods and amending Regulation (EC) No 258/97/EC, COM(2007) 872 final, 2008/0002 (COD).

Commission of the European Communities (2008b), Proposal for a Regulation of the European Parliament and of the Council amending, as regards pharmacovigilance of medicinal products for human use, Regulation (EC) No 726/2004 laying down Community procedures for the authorisation and supervision of medicinal products for human and veterinary use and establishing a European Medicines Agency, COM(2008) 664 final, 2008/0257 (COD).

Commission Recommendation 97/618/EC of 29 July 1997 concerning the scientific aspects and the presentation of information necessary to support applications for the placing on the market of novel foods and novel food ingredients and the preparation of initial assessment reports under Regulation (EC) No 258/97 of the European Parliament and of the Council, *Official Journal L 253*, 16 September 1997, pp. 0001–0036.

Coppens, P., Da Silva, M.F. and Pettman, S. (2006), 'European regulations on nutraceuticals, dietary supplements and functional foods: A framework based on safety', *Toxicology*, **221**(1), 59–74.

Corlett, D. A. and Stier, R. F. (1991), 'Risk assessment within the HACCP system', *Food Control*, **2**(2), 71–72.

Cormier, R. J., Mallet, M., Chiasson, S., Magnússon, H. and Valdimarsson, G. (2007), 'Effectiveness and performance of HACCP-based programs', *Food Control*, **18**(6), 665–671.

Council of Ministers of the European Union (2001), Proposal for a Regulation of the European Parliament and the Council laying down the general principles and requirements of food law, establishing the European Food Authority and laying down procedures in matters of food safety, 8363/01, Brussels.

Courtial, J. P. (1994), 'A coword analysis of scientometrics', *Scientometrics*, **31**(3), 251–260.

Cozzens, S. E. and Woodhouse, E. J. (1995), 'Science, government and the politics of knowledge', in S. Jasanoff, G. Markle, J. Petersen and T. Pinch (eds), *Handbook of Science and Technology Studies*, London: Sage, pp. 533–571.

Crane, D. (1969), 'Social structure in a group of scientists: A test of the invisible college hypothesis', *American Sociological Review*, **34**(3), 335–352.

Crane, D. (1972), *Invisible Colleges. Diffusion of Knowledge in Scientific Communities*, Chicago & London: University of Chicago Press.

Crane, D. (1981), 'Alternative models of ISPAs', in W. M. Evan (ed.), *Knowledge and Power in a Global Society*, Beverley Hills, CA: Sage, pp. 29–47.

Czarniawska, B. and Joerges, B. (1996), 'Travels of ideas', in B. Czarniawska and G. Sevon (eds), *Translating Organizational Change*, Berlin: de Gruyter, pp. 13–48.

Daburon-Garcia, C. (2001), *Le médicament*, Paris: Les Études Hospitalières.

Daemmrich, A. (2004), *Pharmacopolitics: Drug regulation in the United States and Germany*, Chapel Hill: University of North Carolina Press.

Dally, A. (1998), 'Thalidomide: Was the tragedy preventable?', *The Lancet*, **351**(9110), 1197–1199.

Dangoumau, J. (2002), 'Origines de la pharmacologie clinique en France', *Thérapie*, **57**(1), 6–26.

Dawson, P. S. (1992), 'Control of Salmonella in poultry in Great Britain', *International Journal of Food Microbiology*, **15**(3–4), 215–217.

Dawson, R. J. (1995), 'The role of the Codex Alimentarius Commission in setting food standards and the SPS agreement implementation', *Food Control*, **6**(5), 261–265.

Dayan, A. (2000), 'Future problems requiring scientific consideration', *Food and Chemical Toxicology*, **38**(Suppl. 1), S101–106.

Debure, A. (2008), 'Risk analysis principles: The framework binding scientific expertise, food safety policy and international trade', Paper presented at 4S conference, Rotterdam.

Demortain, D. (2008), 'Institutional polymorphism. The designing of the European Food Safety Authority with regard to the European

Medicines Agency', Discussion Paper 50, ESRC Centre for Analysis of Risk and Regulation, April 2008.

Demortain, D. (2009), 'Standards of scientific advice. Risk analysis and the formation of the European Food Safety Authority', in J. Lentsch and P. Weingart (eds), *Scientific Advice to Policy Making: International Comparison*, Berlin: Barbara Budrich, pp. 145–160.

Demortain, D. (2010), 'Regulatory toxicology in controversy. The contentious application of the 90-day rat feeding study to GM safety assessment', Paper given at the conference on Carcinogens, Mutagens, Reproductive Toxicants: The Politics of Limit Values and Low Doses in the Twentieth and Twenty-First Centuries. Strasbourg: IRIST, University of Strasbourg.

Desrosières, A. (2002), *The Politics of Large Numbers. A History of Statistical Reasoning*, Cambridge, MA: Harvard University Press.

DiMasi, J. A. and Grabowski, H. G. (2007), 'The cost of biopharmaceutical R&D: Is biotech different?', *Managerial and Decision Economics*, **28**(4–5), 469–479.

DiMasi, J. A., Hansen, R. W. and Grabowski, H. G. (2003), 'The price of innovation: New estimates of drug development costs', *Journal of Health Economics*, **22**(2), 151–185.

Directive 2001/83/EC of the European Parliament and of the Council of 6 November 2001 on the Community code relating to medicinal products for human use, OJ L 311, 28.11.2001, pp. 67–128.

Directive 2003/53/EC of the European Parliament and of the Council of 18 June 2003 amending for the 26th time Council Directive 76/769/EEC relating to restrictions on the marketing and use of certain dangerous substances and preparations, OJ L 178, 17.7.2003, pp. 24–27.

Directive 2003/99/EC of the European Parliament and of the Council of 17 November 2003 on the monitoring of zoonoses and zoonotic agents, amending Council Decision 90/424/EEC and repealing Council Directive 92/117/EEC, OJ L 325, 12.12.2003, pp. 31–40.

Directive 2004/27/EC of the European Parliament and of the Council of 31 March 2004 amending Directive 2001/83/EC on the Community code relating to medicinal products for human use, OJ L 136, 30.4.2004, pp. 34–57.

Directive 64/433/EEC of the Council of 26 June 1964 on health problems affecting intra-Community trade in fresh meat, OJ 121, 29.7.1964, pp. 2012–2032.

Directive 64/54/EEC of 5 November 1963 on the approximation of the laws of the Member States concerning the preservatives authorized for use in foodstuffs intended for human consumption, OJ 12, 27.1.1964, pp. 161–165.

Directive 71/118/EEC of the Council of 15 February 1971 on health problems affecting trade in fresh poultrymeat, OJ L 55, 8.3.1971, pp. 23–39.

Directive 80/777/EEC of the Council of 15 July 1980 on the approximation of the laws of the Member States relating to the exploitation and marketing of natural mineral waters, OJ L 229, 30.8.1980, pp. 1–10.

Directive 89/107/EEC of the Council of 21 December 1988 on the approximation of the laws of the Member States concerning food additives authorized for use in foodstuffs intended for human consumption, OJ L 40, 11.2.1989, pp. 27–33.

Directive 89/437/EEC of the Council of 20 June 1989 on hygiene and health problems affecting the production and the placing on the market of egg products, OJ L 212, 22.7.1989, pp. 87–100.

Directive 91/492/EEC of the Council of 15 July 1991 laying down the health conditions for the production and the placing on the market of live bivalve molluscs, OJ L 268, 24.9.1991, pp. 1–14.

Directive 91/493/EEC of the Council of 22 July 1991 laying down the health conditions for the production and the placing on the market of fishery products, OJ L 268, 24.9.1991, pp. 15–34.

Directive 92/117/EEC of the Council of 17 December 1992 concerning measures for protection against specified zoonoses and specified zoonotic agents in animals and products of animal origin in order to prevent outbreaks of food-borne infections and intoxications, OJ L 62, 15.3.1993, pp. 38–48.

Directive 92/46/EEC of the Council of 16 June 1992 laying down the health rules for the production and placing on the market of raw milk, heat-treated milk and milk-based products, OJ L 268, 14.9.1992, pp. 1–32.

Directive 93/43/EEC of the Council of 14 June 1993 on the hygiene of foodstuffs, OJ L 175, 19.7.1993, pp. 1–11.

Directive 94/65/EC of the Council of 14 December 1994 laying down the requirements for the production and placing on the market of minced meat and meat preparations, OJ L 368, 31.12.1994, pp. 10–31.

Directive 95/2/EC of the Council of 20 February 1995 on food additives other than colours and sweeteners, OJ L 61, 18.3.1995, pp. 1–40.

Djelic, M. L. (2001), *Exporting the American Model: The Post-War Transformation of European Business*, New York: Oxford University Press.

Djelic, M. L. (2004), 'Social networks and country-to-country transfer: Dense and weak ties in the diffusion of knowledge', *Socio-Economic Review*, **2**, 341–370.

Djelic, M. L. and Quack, S. (2010), *Transnational Communities: Shaping Global Economic Governance*, Cambridge: Cambridge University Press.

Djelic, M.-L. and Sahlin-Andersson, K. (2006), 'A world of governance: The rise of transnational regulation', in M.-L. Djelic and K. Sahlin-Andersson (eds), *Transnational Governance: Institutional Dynamics of Regulation*, Cambridge: Cambridge University Press, pp. 1–28.

Dobson, P. W., Waterson, M. and Davies, S. W. (2003), 'The patterns and implications of increasing concentration in European food retailing', *Journal of Agricultural Economics*, **54**(1), 111–125.

Doern, G. B. and Reed, T. (2001), 'Science and scientists in regulatory governance: A mezzo-level framework for analysis', *Science and Public Policy*, **28**(3), 195–204.

Doll, R. (2003), 'Fisher and Bradford-Hill: Their personal impact', *International Journal of Epidemiology*, **32**, 322–327.

Douglas, M. (1990), 'Risk as a forensic resource', *Daedalus*, **119**(4), 1–16.

Douglas, M. and Wildavsky, A. (1983), *Risk and Culture: An Essay on the Selection of Technical and Environmental Dangers*, San Francisco: University of California Press.

Dowding, K. (1995), 'Model or metaphor? A critical review of the policy network approach', *Political Studies*, **43**(1), 136–158.

Drake, W. J. and Nicolaïdis, K. (1992), 'Ideas, interests, and institutionalization: Trade in services and the Uruguay Round', *International Organization*, **46**(1), 37–100.

Dratwa, J. (2004), 'Social learning with the precautionary principle at the European Commission and the Codex Alimentarius', in B. Reinalda and B. Verbeek (eds), *Decision Making within International Organizations*, London: Routledge, pp. 215–227.

Drori, G. and Meyer, J. (2006), 'Scientization: Making a world safe for organizing, in M.-L. Djelic and K. Sahlin-Andersson (eds), *Transnational Governance: Institutional Dynamics of Regulation*, Cambridge: Cambridge University Press, pp. 31–52.

Drori, G. S., Jang, Y. S. and Meyer, J. W. (2006), 'Sources of rationalized governance: Cross-national longitudinal analyses, 1985–2002', *Administrative Science Quarterly*, **51**(2), 205–229.

Drori, G. S., Meyer, J. W., Ramirez, F. O. and Schofer, E. (2003), *Science in the Modern World Polity: Institutionalization and Globalization*, Stanford, CA: Stanford University Press.

Dudouet, F. X., Mercier, D. and Vion, A. (2006), 'Politiques internationales de normalisation. Quelques jalons pour la recherche empirique', *Revue française de science politique*, **46**(3), 367–392.

Dufour, B. (1999), 'Technical and economic evaluation method for use in improving infectious animal disease surveillance networks', *Veterinary Research*, **30**(1), 27–38.

Eberlein, B. and Grande, E. (2005), 'Beyond delegation: Transnational regulatory regimes and the EU regulatory state', *Journal of European Public Policy*, **12**(1), 89–112.

Edge, D. O., Lemaine, G., MacLeod, R., Mulkay, M. and Weingart, P. (eds) (1976), *Perspectives on the Emergence of Scientific Disciplines*, Chicago: Mouton Aldine.

Edmond, G. (2004), *Expertise in Regulation and Law*. London: Ashgate Publishing.

Edwards, I. R. (1997), 'Who cares about pharmacovigilance?', *European Journal of Clinical Pharmacology*, **53**(2), 83–88.

EFSA (2004), 'Guidance document of the Scientific Panel on Genetically Modified Organisms for the risk assessment of genetically modified plants and derived food and feed', *The EFSA Journal*, **99**, 1–94.

EFSA (2009), *Annual Activity Report of the European Food Safety Authority for 2009*, Parma: EFSA.

Egan, M. (1998), 'Regulatory strategies, delegation and European market integration', *Journal of European Public Policy*, **5**(3), 485–506.

Ehiri, J. E., Morris, G. P. and McEwen, J. (1995), 'Implementation of HACCP in food businesses: The way ahead', *Food Control*, **6**(6), 341–345.

EMEA (2005), Guideline on Risk Management Systems for Medicinal Products for Human Use. Committee for Medicinal Products for Human Use (CHMP), Doc. Ref. EMEA/CHMP/96268/2005.

EMEA and HMA (2005a), Action Plan to Further Progress the European Risk Management Strategy, Doc. Ref. EMEA/115906/2005/Final.

EMEA and HMA (2005b), Implementation of the Action Plan to Further Progress the European Risk Management Strategy: Rolling Two-Year Work Programme (Mid 2005 – Mid 2007), Doc. Ref. EMEA/372687/2005.

EMEA and HMA (2007), European Risk Management Strategy: Achievements to date, Doc. Ref. EMEA/308167/2007.

Engwall, L. and Sahlin-Andersson, K. (2002), *The Expansion of Management Knowledge: Carriers, Flows, and Sources*, Stanford, CA: Stanford University Press.

Evans, J. B. and Cunlife, P. W. (1987), *Study of Control of Medicines*, London: Crown Copyright.

Everson, M. and Vos, E. (2009), 'The scientification of politics and the politicisation of science', in M. Everson and E. Vos (eds), *Uncertain Risks Regulated*, New York: Routledge Cavendish, pp. 1–18.

Ewald, F. (1986), *L'Etat providence*, Paris: Grasset.

Ewald, F. (1991), 'Insurance and risk', in G. Burchell, C. Gordon and P. Miller (eds), *The Foucault Effect: Studies in Governmentality*, London: Harvester Wheatsheaf, pp. 197–210.

Eyck, T. A. T., Thede, D., Bode, G. and Bourquin, L. (2006), 'Is HACCP nothing? A disjoint constitution between inspectors, processors, and consumers and the cider industry in Michigan', *Agriculture and Human Values*, **23**(2), 205–214.

Faden, L. B. and Milne, C. P. (2008), 'Pharmacovigilance activities in the United States, European Union and Japan: Harmonic convergence or convergent evolution?', *Food and Drug Law Journal*, **63**(3), 683–700.

FAO/WHO (2003), *Assuring Food Safety and Quality. Guidelines for Strengthening National Food Control Systems* (Vol. 76), Rome: FAO/WHO.

FDA (1999), 'Managing the risks from medical product use: Creating a risk management framework', Report to the FDA commissioner from the task force on risk management, U.S. Department of Health and Human Services, Food and Drug Administration, May 1999.

Feick, J. (2004), 'Learning and interest accommodation in policy and institutional change: EC risk regulation in the pharmaceuticals sector', *CARR Discussion Paper 25*.

Fischer, F. (2009), *Democracy and Expertise: Reorienting Policy Inquiry*, Oxford: Oxford University Press.

Fischhoff, B., Slovic, P., Lichtenstein, S., Read, S. and Combs, B. (1978), 'How safe is safe enough? A psychometric study of attitudes towards technological risks and benefits', *Policy Sciences*, **9**(2), 127–152.

Fisher, E. (2007), *Risk Regulation and Administrative Constitutionalism*, Oxford: Hart Publishing.

Fleck, L. and Trenn, T. J. (1981), *Genesis and Development of a Scientific Fact*, Chicago: University of Chicago Press.

Fourcade, M. (2006), 'The construction of a global profession: The transnationalization of economics', *American Journal of Sociology*, **112**(1), 145–194.

Freidson, E. (2001), *Professionalism: The Third Logic*, Bristol: Policy Press.

French, M. and Phillips, J. (2000), *Cheated Not Poisoned? Food Regulation in the United Kingdom, 1875–1938*, Manchester: Manchester University Press.

Friedman, M. A., Woodcock, J., Lumpkin, M. M., Shuren, J. E., Hass, A. E. and Thompson, L. J. (1999), 'The safety of newly approved medicines: Do recent market removals mean there is a problem?', *Journal of the American Medical Association*, **281**(18), 1728–1734.

Fritsch, P. (1985), 'Situations d'expertise et "expert-système"', Actes du colloque *Situations d'expertise et socialisation des savoirs,* Saint-Etienne: CRESAL.

Fulponi, L. (2006), 'Private voluntary standards in the food system: The perspective of major food retailers in OECD countries', *Food Policy*, **31**(1), 1–13.

Funtowicz, S. O. and Ravetz, J. R. (1992), 'Three types of risk assessment and the emergence of post-normal science', in S. Krimsky and D. Golding (eds), *Social Theories of Risk*, Westport, CT: Praeger, pp. 251–273.

Gabe, J. (1995), 'Health, medicine and risk: The need for a sociological approach', in J. Gabe (ed.), *Medicine, Health and Risk: Sociological Approaches*, Oxford: Blackwell Publishers, pp. 1–17.

Galli, C. L., Marinovich, M. and Lotti, M. (2008), 'Is the acceptable daily intake as presently used an axiom or a dogma?', *Toxicology Letters*, **180**(2), 93–99.

Gallo, M. A. (1996), 'History and scope of toxicology', in C. D. Klassen (ed.), *Casarett and Doull's Toxicology: The Basic Science of Poisons*, New York: McGraw Hill, pp. 3–11.

Garland, D. (2003), 'The rise of risk', in R. Ericson and A. Doyle (eds), *Risk and Morality*, Toronto: Toronto University Press, pp. 48–86.

Garrett, E. S., Jahncke, M. L. and Cole, E. A. (1998), 'Effects of Codex and GATT', *Food Control*, **9**(2–3), 177–182.

Gehring, T. and Krapohl, S. (2007), 'Supranational regulatory agencies between independence and control: The EMEA and the authorization of pharmaceuticals in the European Single Market', *Journal of European Public Policy*, **14**(2), 208–226.

Giddens, A. (1999), 'Risk and responsibility', *The Modern Law Review*, **62**(1), 1–10.

Gilbert, J. and Scott, P. (2000), 'Editorial', *Food Additives and Contaminants*, **17**(1), 1.

Gilpin, R. and Wright, C. (eds) (1964), *Scientists and National Policy-Making*, New York: Columbia University Press.

Gold, H. (1971), 'The American College of Clinical Pharmacology', *Journal of Clinical Pharmacology*, **11**(5), 321–322.

Goldstein, B. D. (1990), 'The problem with the margin of safety: Toward the concept of protection', *Risk Analysis*, **10**(1), 7–10.

Granjou, C. and Barbier, M. (2009), *Métamorphoses de l'expertise. Précaution et maladies à prions*, Versailles: Quae.

Graz, J. C. (2003), 'How powerful are transnational elite clubs? The social myth of the World Economic Forum', *New Political Economy*, **8**(3), 321–340.

Griffith, C. J. (2006), 'Food safety: where from and where to?', *British Food Journal*, **108**(1), 6–15.

Groenewegen, P. (1991), 'The construction of expert advice on health risks', *Social Studies of Science*, **21**(2), 257–278.

Groenewegen, P. (2002), 'Accommodating science to external demands: The emergence of Dutch toxicology', *Science, Technology & Human Values*, **27**(4), 479.

Groenleer, M. (2009), *The Autonomy of European Union Agencies. A Comparative Study of Institutional Development*, Delft: Eburon Uitgeverij BV.

Gusfield, J. (1975), *Community: A Critical Response*, Oxford: Basil Blackwell.

Guston, D. H. (2001), 'Boundary organizations in environmental policy and science: An introduction', *Science, Technology and Human Values*, **26**(4), 399–408.

Haas, P. M. (1989), 'Do regimes matter – epistemic communities and Mediterranean pollution-control', *International Organization*, **43**(3), 377–403.

Haas, P. M. (1992), 'Introduction: Epistemic communities and international policy coordination', *International Organization*, **46**(1), 1–35.

Haas, P. M. (2000), 'International institutions and social learning in the management of global environmental risks', *Policy Studies Journal*, **28**(3), 558–575.

Haberer, J. (1969), *Politics and the Community of Science*, New York: Van Nostrand Reinhold Co.

Hancher, L. (1990), *Regulating for Competition: Government, Law, and the Pharmaceutical Industry in the United Kingdom and France*, Oxford: Clarendon Press.

Hancher, L. and Moran, M. (1989), 'Organizing regulatory space', in L. Hancher and M. Moran (eds), *Capitalism, Culture and Regulation*, Oxford: Clarendon Press, pp. 271–299.

Hansson, S. O. (2010), 'Risk: Objective or subjective, facts or values', *Journal of Risk Research*, **13**(2), 231–238.

Harrigan, W. F. (1993), 'The ISO 9000 series and its implications for HACCP', *Food Control*, **4**(2), 105–111.

Harris, R. A. and Milkis, S. M. (1996), *The Politics of Regulatory Change: A Tale of Two Agencies*, New York: Oxford University Press.

Hartford, C. (2006), 'An industry perspective on clinical drug safety. Risk management: From concept to practice', *Canadian Journal of Clinical Pharmacology*, **13**(1), e22–e28.

Hasenclever, A., Mayer, P. and Rittberger, V. (1996), 'Interests, power, knowledge: The study of international regimes', *International Studies Quarterly*, **40**, 177–228.

Hathaway, S. C. (1995), 'Harmonization of international requirements under HACCP-based food control systems', *Food Control*, **6**, 267–276.

Hauray, B. (2005), 'Politique et expertise scientifique. La régulation Européenne des médicaments', *Sociologie du Travail*, **47**(1), 57–75.

Hauray, B. (2006), *L'Europe du médicament: Politique, expertise, intérêts privés*, Paris: Presses de Sciences Po.

Hauray, B. and Urfalino, P. (2009), 'Mutual transformation and the development of European policy spaces. The case of medicines licensing', *Journal of European Public Policy*, **16**(3), 431–449.

Havinga, T. (2006), 'Private regulation of food safety by supermarkets', *Law and Policy*, **28**(4), 515–533.

Hawkins, R. W. (1995), 'Standards-making as technological diplomacy: Assessing objectives and methodologies in standards institutions', in R. Hawkins, R. Mansell and J. Skea (eds), *Standards, Innovation, and Competitiveness: The Politics and Economics of Standards in Natural and Technical Environments*, Aldershot, UK and Brookfield, VT, USA: Edward Elgar Publishing, pp. 147–156.

Hayes, M. V. (1992), 'On the epistemology of risk: Language, logic and social science', *Social Science & Medicine*, **35**(4), 401–407.

Heclo, H. (1978), 'Issue networks and the executive establishment', in A. King (ed.), *The New American Political System*, Washington, DC: American Enterprise Institute, pp. 87–124.

Henson, S. and Caswell, J. (1999), 'Food safety regulation: An overview of contemporary issues', *Food Policy*, **24**(6), 589–603.

Hepburn, P., Howlett, J., Boeing, H., Cockburn, A., Constable, A., Davi, A., et al. (2008), 'The application of post-market monitoring to novel foods', *Food and Chemical Toxicology*, **46**(1), 9–33.

Hilgartner, S. (1992), 'The social construction of risk objects', in J. F. Short and L. Clarke (eds), *Organizations, Uncertainties, and Risk*, Boulder, CO: Westview Press, pp. 40–53.

Hill, A. B. (1965), 'The environment and disease: Association or causation?', *Proceedings of the Royal Society of Medicine*, **58**, 295–300.

Hlywka, J. J., Reid, J. E. and Munro, I. C. (2003), 'The use of consumption data to assess exposure to biotechnology-derived foods and the feasibility of identifying effects on human health through post-market monitoring', *Food and Chemical Toxicology*, **41**(10), 1273–1282.

HoA (2003), Establishing a European Risk Management Strategy: Heads of Agencies Ad Hoc Working Group, MCA/PL/JM/HoA/Summary Report.

Hoch, P. K. (1987), 'Institutional versus intellectual migrations in the nucleation of new scientific specialties', *Studies in the History and Philosophy of Science*, **18**(4), 481–500.

Hofmann, J. (1995), 'Implicit theories in policy discourse: An inquiry into the interpretations of reality in German technology policy', *Policy Sciences*, **28**(2), 127–148.

Holland, D. and Pope, H. (2004), *EU Food Law and Policy*, The Hague: Kluwer Law International.

Holt, G. and Henson, S. (2000), 'Quality assurance management in small meat manufacturers', *Food Control*, **11**(4), 319–326.

Hood, C. (1983), *The Tools of Government*, Basingstoke: Macmillan.

Hood, C. and Jones, D. K. C. (1996), *Accident and Design: Contemporary Debates in Risk Management*, London: Routledge.

Hood, C., Rothstein, H. and Baldwin, R. (2001), *The Government of Risk: Understanding Risk Regulation Regimes*, Oxford: Oxford University Press.

Horton, L. R. (2001), 'Risk analysis and the law: International law, the World Trade Organization, Codex Alimentarius and national legislation, *Food Additives & Contaminants*, **18**(12), 1057–1067.

Howlett, J., Edwards, D. G., Cockburn, A., Hepburn, P., Kleiner, J., Knorr, D., et al. (2003), 'The safety assessment of Novel Foods and concepts to determine their safety in use', *International Journal of Food Sciences and Nutrition*, **54**, 1–32.

Hubscher, R. (1999), *Les maîtres des bêtes: Les vétérinaires dans la société française (XVIIIe–XXe siècle)*, Paris: Odile Jacob.

Hülsse, R. and Kerwer, D. (2007), 'Global standards in action: insights from anti-money laundering regulation', *Organization*, **14**(5), 625–642.

Hutter, B. M. (2001), *Regulation and Risk: Occupational Health and Safety on the Railways*, Oxford: Oxford University Press.

Hutter, B. M. (2006), 'Risk, regulation, and management', in P. Taylor-Gooby and J. Zinn (eds), *Risk in Social Science*, Oxford: Oxford University Press, pp. 202–227.

Hyslop, D. L. (2002), 'Pharmaceutical risk management: A call to arms for pharmacoepidemiology', *Pharmacoepidemiology and Drug Safety*, **11**, 417–420.

ICH (2002), Final Concept Paper E2E: Pharmacovigilance Planning, Dated and endorsed by the Steering Committee on 11 September 2002, International Conference on Harmonisation of Technical Requirements for Registration of Pharmaceuticals for Human Use.

ICH (2004), ICH E2E: Pharmacovigilance Planning (PvP), Current *Step 4* Version, 18 November 2004.

ICMSF (1988), *Microorganisms in Foods 4. Application of the Hazard Analysis Critical Control Point (HACCP) System to Ensure Microbiological Safety and Quality*, Oxford: Blackwell Scientific Publications.

ILSI Europe (1993), *A Simple Guide to Understanding and Applying the Hazard Analysis Critical Control Point Concept*, Brussels: ILSI Europe.

Inman, B. (1993), '30 years in postmarketing surveillance. A personal perspective', *Pharmacoepidemiology and Drug Safety*, **2**(4–5), 239–258.

Irwin, A., Rothstein, H., Yearley, S. and McCarthy, E. (1997), 'Regulatory science – Towards a sociological framework', *Futures*, **29**(1), 17–31.

ISPE (2003), 'Risk assessment of observational data: Good pharmacovigilance practices and pharmacoepidemiologic assessment', Communication at the Risk Management Public Workshop, Washington, DC.

Jacobs, S. (1987), 'Scientific community: Formulations and critique of a sociological motif', *British Journal of Sociology*, **38**(2), 266–276.

Jacobsson, B. (2000), 'Standardization and expert knowledge', in N. Brunsson and B. Jacobsson (eds), *A World of Standards*, Oxford: Oxford University Press, pp. 127–137.

Jahnke, M. and Kühn, K. D. (2003), 'Use of the hazard analysis and critical control points (HACCP) risk assessment on a medical device for parenteral application', *PDA Journal of Pharmaceutical Science and Technology*, **57**(1), 32–42.

James, O. and Lodge, M. (2003), 'The limitations of "policy transfer" and "lesson drawing" for public policy research', *Political Studies Review*, **1**(2), 179–193.

Jas, N. (2010), 'To protect mankind from the danger of the chemical age? International experts, international organisations in the development of international expertise on food additives, 1954–1963', Paper given at the conference on Carcinogens, Mutagens, Reproductive Toxicants: The Politics of Limit Values and Low Doses in the Twentieth and Twenty-First Centuries, Strasbourg: IRIST, University of Strasbourg.

Jasanoff, S. (1987), 'Contested boundaries in policy-relevant science', *Social Studies of Science*, **17**(2), 195–230.

Jasanoff, S. (1990a), *The Fifth Branch: Science Advisers as Policymakers*, Cambridge, MA: Harvard University Press.

Jasanoff, S. (1990b), 'American exceptionalism and the political acknowledgment of risk', *Daedalus*, **119**(4), 61–81.

Jasanoff, S. (1992), 'Science, politics, and the renegotiation of expertise at EPA', *Osiris*, **7**, 194–217.

Jasanoff, S. (1995a), *Science at the Bar*, Cambridge, MA: Harvard University Press.

Jasanoff, S. (1995b), 'Procedural choices in regulatory science', *Technology in Society*, **17**(3), 279–293.

Jasanoff, S. (2003), 'Breaking the waves in science studies: Comment on Harry M. Collins and Robert Evans, "the third wave of science studies"', *Social Studies of Science*, **33**(3), 389–400.

Jasanoff, S. (2005), *Designs on Nature: Science and Democracy in Europe and the United States*, Princeton, NJ: Princeton University Press.

Johnson, B. B. and Covello, V. T. (1987), *The Social and Cultural Construction of Risk: Essays on Risk Selection and Perception*, Dordrecht: Reidel.

Joly, P. B. (2007), 'Scientific expertise in public arenas: Lessons from the French experience', *Journal of Risk Research*, **10**(7), 905–924.

Joly, P.-B. and Barbier, M. (2001), 'Que faire des désaccords entre comités d'experts? Les leçons de la guerre du bœuf', *Revue Risques,* **47**, 87–94.

Jonas, D. A., Antignac, E., Antoine, J. M., Classen, H. G., Huggett, A., Knudsen, I., et al. (1996), 'The safety assessment of novel foods: Guidelines prepared by ILSI Europe Novel Food Task Force', *Food and Chemical Toxicology*, **34**(10), 931–940.

Jones, C. O. (1970), *An Introduction to the Study of Public Policy*, Belmont, CA: Duxbury Press.

Jongeneel, S. and van Schothorst, M. (1992), 'HACCP, product liability and due diligence', *Food Control*, **3**(3), 122–126.

Jongeneel, S. and van Schothorst, M. (1994), 'Line monitoring, HACCP and food safety', *Food Control*, **5**(2), 107–110.

Jouve, J. L. (1994a), 'European standards for quality management', *Food Control*, **5**(4), 227–229.

Jouve, J. L. (1994b), 'HACCP as applied in the EEC', *Food Control*, **5**(3), 181–186.

Jouve, J.-L. (1999), *An Overall Strategy for Risk Assessment of Microbiological Hazards, Risk Assessment of Microbiological Hazards in Food: A Joint FAO/WHO Expert Consultation*, Geneva: WHO.

Jouve, J. L., Stringer, M. F. and Baird-Parker, A. C. (1998), *Food Safety Management Tools*, Washington, DC: ILSI Press.

JRC (2008), *The 2008 EU Industrial R&D Investment Scoreboard*, Ispra: JRC/DG RTD, European Commission.

Jukes, D. J. (1993), *Food Legislation in the UK: A Concise Guide*, Oxford and Boston: Butterworth-Heinemann.

Käferstein, F. K., Motarjemi, Y. and Bettcher, D. W. (1997), 'Foodborne disease control: A transnational challenge', *Emerging Infectious Diseases*, **3**(4), 503–510.

Keck, M. E. and Sikkink, K. (1998), *Activists beyond Borders: Advocacy Networks in International Politics*, Ithaca, NY: Cornell University Press.

Keeler, J. T. S. (1993), 'Opening the window for reform: mandates, crises, and extraordinary policy-making', *Comparative Political Studies*, **25**(4), 433–486.

Keohane, R. O. and Nye, J. S. (1977), *Power and Interdependence: World Politics in Transition*, Boston: Little Brown.

Kerwer, D. (2005), 'Rules that many use: Standards and global regulation', *Governance*, **18**(4), 611–632.

Khandke, S. S. and Mayes, T. (1998), 'HACCP implementation: A practical guide to the implementation of the HACCP plan', *Food Control*, **9**(2–3), 103–109.

Kingdon, J. W. (1984), *Agendas, Alternatives and Public Policy*, Boston: Little Brown.

Knorr-Cetina, K. D. (1982), 'Scientific communities or trans-epistemic arenas of research – A critique of quasi-economic models of science', *Social Studies of Science*, **12**(1), 101–130.

Knowles, M. and Walker, R. (1984), 'Editorial', *Food Additives and Contaminants*, **1**(1), 1–2.

Knowles, M. E., Bell, J. R., Norman, J. A. and Watson, D. H. (1991), 'Surveillance of potentially hazardous chemicals in food in the United Kingdom', *Food Additives and Contaminants*, **8**(5), 551–564.

König, A., Cockburn, A., Crevel, R. W. R., Debruyne, E., Grafstroem, R., Hammerling, U., et al. (2004), 'Assessment of the safety of foods derived from genetically modified (GM) crops', *Food and Chemical Toxicology*, **42**(7), 1047–1088.

Krapohl, S. (2008), *Risk Regulation in the Single Market: The Governance of Pharmaceuticals and Foodstuffs in the European Union*, London: Palgrave Macmillan.

Kreher, A. (1997), 'Agencies in the European Community – A step towards administrative integration in Europe', *Journal of European Public Policy*, **4**(2), 225–245.

Kuhn, T. S. (1970), *The Structure of Scientific Revolutions*, 2nd edn, Chicago: University of Chicago Press.

Kurunmäki, L. (2004), 'A hybrid profession – The acquisition of management accounting expertise by medical professionals', *Accounting, Organizations and Society*, **29**(3–4), 327–347.

Kvenberg, J. E. (1998), 'Introduction to food safety HACCP', *Food Control*, **9**(2/3), 73–74.

Kwak, N. S. and Jukes, D. J. (2001), 'Functional foods. Part 2: The impact on current regulatory terminology', *Food Control*, **12**(2), 109–117.

Lachance, P. (1997), 'How HACCP started', *Food Technology*, **51**(5), 35.

Lakoff, A. (2007), 'The right patients for the drug: Managing the placebo effect in antidepressant trials', *BioSocieties*, **2**(1), 57–71.

Langlitz, N. (2009), 'Pharmacovigilance and post-black market surveillance', *Social Studies of Science*, **39**(3), 395–420.

Latour, B. (1987), *Science in Action: How to Follow Scientists and Engineers through Society*, Cambridge, MA: Harvard University Press.

Latour, B. (1999), *Politiques de la nature: comment faire entrer les sciences en démocratie*, Paris: La Découverte.

Lazarou, J., Pomeranz, B. H. and Corey P. N. (1998), 'Incidence of adverse drug reactions in hospitalized patients. A meta-analysis of prospective studies', *Journal of the American Medical Association*, **279**, 1200–1205.

Lea, L. and Hepburn, P. (2002), *Cholesterol Lowering Vegetable Oil Spreads: Results of a Post Launch Monitoring Programme, Separata of Deutsche Forschungesgemeinshaft, Symposium on Function Foods: Safety Aspects*, Weinheim: Wiley-VCH.

Lechat, P. and Fontagne, J. (1973), 'Présentation d'une nouvelle rubrique de *"Thérapie"* consacrée aux informations sur les effets indésirables des médicaments', *Thérapie*, **28**(1), 169–171.

Lemaine, G., MacLeod, R., Mulkay, M. and Weingart, P. (eds) (1976), *Perspectives on the Emergence of Scientific Disciplines*, The Hague, Paris and Chicago: Mouton and Aldine.

Lenz, W. (1992), 'The history of thalidomide', UNITH Congress, available at www.thalidomidesociety.co.uk/publications.htm.

Levidow, L., Murphy, J. and Carr, S. (2007), 'Recasting "substantial equivalence": Transatlantic governance of GM food', *Science, Technology & Human Values*, **32**(1), 26–64.

Levi-Faur, D. (2005), 'The global diffusion of regulatory capitalism', *Annals of the American Academy of Political and Social Science*, **598**, 12–32.

Leydesdorff, L. (1989), 'Words and co-words as indicators of intellectual organization', *Research Policy*, **18**(4), 209–223.

Lezaun, J. (2006), 'Creating a new object of government: Making genetically modified organisms traceable', *Social Studies of Science*, **36**(4), 499.

Lievrouw, L. A. (1989), 'The invisible college reconsidered – Bibliometrics and the development of scientific communication-theory', *Communication Research*, **16**(5), 615–628.

Lindquist, M. (2003), *Seeing and Observing in International Pharmacovigilance: Achievements and Prospects in Worldwide Drug Safety*, Uppsala: Uppsala Monitoring Centre.

Lindsay, D. G. (1987), 'Estimation of the dietary intake of chemicals in food', *Food Additives & Contaminants*, **3**(1), 71–88.

Linnerooth-Bayer, J., Löfstedt, R. E. and Sjöstedt, G. (eds) (2001), *Transboundary Risk Management*, London: Earthscan Publications.

Lowrance, W. W. (1976), *Of Acceptable Risk: Science and the Determination of Safety*, Los Altos, CA: William Kaufman.

Lu, F. C. (1988), 'Acceptable daily intake: Inception, evolution, and application', *Regulatory Toxicology and Pharmacology*, **8**(1), 45–60.

Luhmann, N. (1993), *Risk: A Sociological Theory*, New York: de Gruyter.

Maasen, S. and Weingart, P. (2005), *Democratization of Expertise?*

Exploring Novel Forms of Scientific Advice in Political Decision-Making, The Hague: Kluwer Academic Publishing.

MacKenzie, D. (2008), 'Producing accounts: Finitism, technology and rule-following', in M. Mazzotti (ed.), *Knowledge as Social Order: Rethinking the Sociology of Barry Barnes*, Aldershot, UK and Burlington, VT, USA: Ashgate, pp. 99–117.

MacLeod, R. M. (ed.) (1988), *Government and Expertise: Specialists, Administrators and Professionals, 1860–1919*, Cambridge: Cambridge University Press.

Macmaoláin, C. (2007), *EU Food Law: Protecting Consumers and Health in a Common Market*, Oxford & Portland, OR: Hart Publishing.

Maennl, U. (2008), 'Pharmacovigilance: A company-wide challenge', *Applied Clinical Trials*, **17**, 50–58.

Majone, G. (1989), *Evidence, Argument, and Persuasion in the Policy Process*, New Haven, CT: Yale University Press.

Majone, G. (1994), 'The rise of the regulatory state in Europe', *West European Politics*, **17**(3), 77–101.

Majone, G. (1997), 'The new European agencies: Regulation by information', *Journal of European Public Policy*, **4**(2), 262–275.

Majone, G. (1999), 'The credibility crisis of community regulation', *Journal of Common Market Studies*, **38**(2), 273–302.

Malaspina, A. (1984), *ILSI. In Cent cinquantième anniversaire de la création de la Chaire de Toxicologie de la Faculté de Pharmacie de Paris et Jubilé scientifique du Professeur René Truhaut*, Paris: Comité d'organisation du jubilé scientifique du Prof. R. Truhaut, pp. 681–682.

Marcussen, M. (2006), 'The transnational governance network of central bankers', in M.-L. Djelic and K. Sahlin-Andersson (eds), *Transnational Governance: Institutional Dynamics of Regulation*, Oxford: Oxford University Press, pp. 180–204.

Marier, P. (2008), 'Empowering epistemic communities: Specialised politicians, policy experts and policy reform', *West European Politics*, **31**(3), 513–533.

Marin, B. and Mayntz, R. (1991), *Policy Networks: Empirical Evidence and Theoretical Considerations*, Frankfurt: Campus Verlag.

Marks, H. (2008), 'Making risks visible. The science and politics of adverse drug reactions', in J. P. Gaudillière and V. Hess (eds), *Ways of Regulating: Therapeutic Agents between Plants, Shops and Consulting Rooms*, Berlin: Max Planck Institute für Wissenschaftsgeschichte Preprint Series No. 383, pp. 105–122.

Marks, H. M. (1997), *The Progress of Experiment: Science and Therapeutic Reform in the United States, 1900–1990*, Cambridge: Cambridge University Press.

Marquand, D. (2004), *Decline of the Public: The Hollowing Out of Citizenship*, London: Polity Press.

Marsh, D. and Rhodes, R. A. W. (eds) (1992), *Policy Networks in British Government*, Oxford: Clarendon Press.

Marth, E. H. (2001), 'The emergence of food microbiology: From dairy microbiology to food microbiology', *Dairy, Food and Environmental Sanitation*, **21**, 818–824.

Mattli, W. (2001), 'The politics and economics of international institutional standards setting: An introduction', *Journal of European Public Policy*, **8**(3), 328–344.

Mattli, W. and Büthe, T. (2003), 'Setting international standards', *World Politics*, **56**, 1–42.

Mayes, T. (1992), 'Simple users' guide to the hazard analysis critical control point concept for the control of food microbiological safety', *Food Control*, **3**(1), 14–19.

Mayes, T. (1998), 'Risk analysis in HACCP: Burden or benefit?', *Food Control*, **9**(2–3), 171–176.

Mayntz, R. (2010), 'Global structures: markets, organizations, networks – and communities?', in M. L. Djelic and S. Quack (eds), *Transnational Communities: Shaping Global Economic Governance*, Cambridge: Cambridge University Press, pp. 37–54.

Mayntz, R. and Scharpf, F. W. (1975), *Policy-Making in the German Federal Bureaucracy*, Amsterdam: Elsevier.

McGoey, L. (2007), 'On the will to ignorance in bureaucracy', *Economy and Society*, **36**(2), 212–235.

Mercer, D. (2004), 'Hyper-experts and the vertical integration of expertise in EMF/RF litigation', in G. Edmond (ed.), *Expertise in Regulation and Law*, Aldershot: Ashgate, pp. 85-97.

Merton, R. K. and Storer, N. W. (1973), *The Sociology of Science: Theoretical and Empirical Investigations*, Chicago: University of Chicago Press.

Merz, M. (2006), 'Différenciation interne des sciences. Constructions discursives et pratiques épistémiques autour de la simulation', in J.-Ph. Leresche, M. Benninghoff, F. Crettaz von Roten and M. Merz (eds), *La fabrique des sciences: Des institutions aux pratiques*, Genève: PPUR, pp. 165–182.

Meyer, J. W. (1994), 'Rationalized environments', in J. W. Meyer and R. Scott (eds), *Institutional Environments and Organizations: Structural Complexity and Individualism*, London: Sage Publications, pp. 28–54.

Meyer, J. and Rowan, B. (1977), 'Institutionalized organizations: Formal structure as myth and ceremony', *American Journal of Sociology*, **83**, 340–363.

Meyer, J. W., Boli, J., Thomas, G. and Ramirez, F. (1997), 'World society and the nation-state', *American Journal of Sociology*, **103**(1), 144–181.

Miller, D. (1999), 'Risk, science and policy: Definitional struggles, information management, the media and BSE', *Social Science & Medicine*, **49**(9), 1239–1255.

Miller, P., Kurunmäki, L. and O'Leary, T. (2008), 'Accounting, hybrids and the management of risk', *Accounting, Organizations and Society*, **33**(7–8), 942–967.

Mills, C. W. (1956), *The Power Elite*, New York: Oxford University Press.

Millstone, E. (2007), 'Can food safety policy-making be both scientifically and democratically legitimated? If so, how?', *Journal of Agricultural & Environmental Ethics*, **20**(5), 483–508.

Millstone, E. and van Zwanenberg, P. (2002), 'The evolution of food safety policy-making institutions in the UK, EU and Codex Alimentarius', *Social Policy & Administration*, **36**(6), 593–609.

Millstone, E., Brunner, E. and Mayer, S. (1999), 'Beyond substantial equivalence', *Nature*, **401**(6753), 525–526.

Miremont, G., Haramburu, F., Bégaud, B., Péré, J.C. and Dangoumau, J. (1994), 'Adverse drug reactions: Physician's opinions versus causality assessment method', *European Journal of Clinical Pharmacology*, **6**, 285–292.

Molle, L. (1984), 'Éloge du Professeur René Truhaut', *Revue d'histoire de la pharmacie*, **72**(262), 340–348.

Moran, M. (2002), 'Understanding the regulatory state', *British Journal of Political Science*, **32**(02), 391–413.

Morris, C. (2003), '75 years of food frontiers', *Food Engineering*, 9 September 2003.

Morrison, A. (1982), 'Risk analysis: Is EPA changing the rules?', *Civil Engineering–ASCE*, **52**(10), 60–63.

Moseley, B. (1991), 'Control of novel, including genetically engineered, foods in the United Kingdom', *Food Control*, **1**(2), 199–201.

Mossel, D.A. (1989), 'Adequate protection of the public against food-transmitted diseases of microbial aetiology: Achievements and challenges, half a century after the introduction of the Prescott–Meyer–Wilson strategy of active intervention', *International Journal of Food Microbiology*, **9**(4), 271–294.

Mossel, D. A. (1995), 'Principles of food control and food hygiene in the European single market', *Food Control*, **6**(5), 289–293.

Motarjemi, Y. (2000), 'Regulatory assessment of HACCP: A FAO/WHO consultation on the role of government agencies in assessing HACCP', *Food Control*, **11**(5), 341–344.

Motarjemi, Y. and Käferstein, F. (1999), 'Food safety, hazard analysis and critical control point and the increase in foodborne diseases: A paradox?', *Food Control*, **10**(4–5), 325–333.

Motarjemi, Y., Käferstein, F., Moy, G., Miyagawa, S. and Miyagishima, K. (1996), 'Importance of HACCP for public health and development. The role of the World Health Organization', *Food Control*, **7**(2), 77–85.

Mulkay, M. (1976), 'The mediating role of the scientific elite', *Social Studies of Science*, **6**(3/4), 445–470.

Mulkay, M. J., Gilbert, G. N. and Woolgar, S. (1975), 'Problem areas and research networks in science', *Sociology*, **9**(2), 187–204.

NACMCF (1991), 'Recommendations of the US National Advisory Committee on Microbiological Criteria for Food. I. HACCP principles, II. Meat and poultry, III. Seafood', *Food Control*, **2**(4), 202–211.

Nelkin, D. (1975), 'The political impact of technical expertise', *Social Studies of Science*, **5**(1), 35–54.

Nelkin, D. (ed.), (1984), *Controversy: Politics of Technical Decisions*, London: Sage Publications.

Nelson, R. (2000), 'We need a postmarketing drug development process!', *Pharmacoepidemiology and Drug Safety*, **9**(3), 253–255.

Nestlé, M. (2002), *Food Politics. How the Food Industry Influences Nutrition and Health*, Berkeley and San Francisco: University of California Press.

Noll, R. G. (1996), 'Reforming risk regulation', *The Annals of the American Academy of Political and Social Science*, **545**, 165–175.

Noma, E. (1984), 'Co-citation analysis and the invisible college', *Journal of the American Society for Information Science*, **35**(1), 29–33.

Notermans, S., Barendsz, A. W. and Rombouts, F. (2002), 'The evolution of microbiological risk assessment in food production', in M. Brown and M. Stringer (eds), *Microbiological Risk Assessment in Food Processing*, Cambridge: Woodhead Publishing & CRC Press.

Notermans, S., Gallhoff, G., Zwietering, M. H. and Mead, G. C. (1995), 'Identification of critical control points in the HACCP system with a quantitative effect on the safety of food products', *Food Microbiology*, **12**(2), 93–98.

Notermans, S., Mead, G. C. and Jouve, J. L. (1996), 'Food products and consumer protection: A conceptual approach and a glossary of terms', *International Journal of Food Microbiology*, **30**(1–2), 175–185.

Notermans, S., Nauta, M. J., Jansen, J., Jouve, J. L. and Mead, G. C. (1998), 'A risk assessment approach to evaluating food safety based on product surveillance', *Food Control*, **9**(4), 217–223.

NRC (1983), *Risk Assessment in the Federal Government: Managing the Process*, Washington, DC: National Academies Press.

NRC (1985a), *Meat and Poultry Inspection: The Scientific Basis of the Nation's Program*, Washington, DC: National Academies Press.

NRC (1985b), *An Evaluation of the Role of Microbiological Criteria for Foods and Food Ingredients*, Washington, DC: National Academies Press.

NRC (1989), *Improving Risk Communication*, Washington, DC: National Academies Press.

NRC (1996), *Understanding Risk: Informing Decisions in a Democratic Society*, Washington, DC: National Academies Press.

Nunn, R. (2008), 'A network model of expertise', *Bulletin of Science, Technology & Society*, **28**(5), 414–427.

OECD (2003), *Report on the Questionnaire on Biomarkers, Research on Safety of Novel Foods and Feasibility of Post-Market Monitoring, Series on the Safety of Novel Foods and Feeds, 8*, ENV/JM/MONO (2003)9. Paris: OECD.

Olson, M. K. (2002), 'Pharmaceutical policy change and the safety of new drugs', *Journal of Law and Economics*, **45**(2), 615–642.

Ong, A. and Collier, S. J. (2005), 'Global assemblages, anthropological problems', in A. Ong and S. J. Collier (eds), *Global Assemblages: Technology, Politics, and Ethics as Anthropological Problems*, Oxford: Blackwell Publishing, pp. 3–21.

O'Rourke, R. (2005), *European Food Law*, London: Sweet & Maxwell.

Otway, H. and Thomas, K. (1982), 'Reflections on risk perception and policy', *Risk Analysis*, **2**(1), 69–82.

Panisello, P. J., Quantick, P. C. and Knowles, M. J. (1999), 'Towards the implementation of HACCP: Results of a UK regional survey', *Food Control*, **10**(2), 87–98.

Panisello, P. J., Rooney, R., Quantick, P. C. and Stanwell-Smith, R. (2000), 'Application of foodborne disease outbreak data in the development and maintenance of HACCP systems', *International Journal of Food Microbiology*, **59**(3), 221–234.

Parmar, I. (2002), 'American foundations and the development of international knowledge networks', *Global Networks*, **2**(1), 13–30.

Paulus, I. L. E. (1974), *The Search for Pure Food: A Sociology of Legislation in Britain*, London: M. Robertson.

Peattie, M. E., Buss, D. H., Lindsay, D. G. and Smart, G. A. (1983), 'Reorganization of the British total diet study for monitoring food constituents from 1981', *Food and Chemical Toxicology*, **21**(4), 503–507.

Perissich, R. (1990), 'Food control in the European Community', *Food Control*, **1**(3), 130–131.

Permanand, G. (2006), *EU Pharmaceutical Regulation: The Politics of Policy-Making*, Manchester: Manchester University Press.

Permanand, G. and Mossialos, E. (2005), 'Constitutional asymmetry and pharmaceutical policy-making in the European Union', *Journal of European Public Policy*, **12**(4), 687–709.

Permanand, G., Mossialos, E. and McKee, M. (2006), 'Regulating medicines in Europe: The European Medicines Agency, marketing authorisation, transparency and pharmacovigilance', *Clinical Medicine*, **6**(1), 87–90.

Perrow, C. (1982), 'Not risk but power', *Contemporary Sociology*, **11**(3), 298–300.

Persson, I. (2003), 'Risk management: A regulatory perspective, II', Presentation in a seminar on Pharmaceutical Risk Management.

Petiteville, F. and Smith, A. (2006), 'Analyser les politiques publiques internationales', *Revue française de science politique*, **56**(3), 357–366.

Pielke, R. A. (2007), *The Honest Broker: Making Sense of Science in Policy and Politics*, Cambridge University Press.

Pirmohamed, M., James, S., Meakin, S., Green, C. and Scott, A. K. (2004), 'Adverse drug reactions as a cause of admission to hospital: Prospective analysis of 18 820 patients', *British Medical Journal*, **329**, 15–19.

Pollak, R. A. (1995), 'Regulating risks', *Journal of Economic Literature*, **33**(1), 179–191.

Pollitt, C., Bathgate, K., Caulfield, J., Smullen, A. and Talbot, C. (2001), 'Agency fever? Analysis of an international policy fashion', *Journal of Comparative Policy Analysis*, **3**(3), 271–290.

Porter, T. M. (1992), 'Quantification and the accounting ideal in science', *Social Studies of Science*, **22**(4), 633–651.

Porter, T. M. (1995), *Trust in Numbers: The Pursuit of Objectivity in Science and Public Life*, Princeton, NJ: Princeton University Press.

Post, D. L. (2006), 'The precautionary principle and risk assessment in international food safety: How the World Trade Organization influences standards', *Risk Analysis*, **26**(5), 1259–1273.

Poulsen, E. (1995), 'René Truhaut and the acceptable daily intake: A personal note', *Teratogenesis, Carcinogenesis, and Mutagenesis*, **15**, 273–275.

Power, M. (2004), *The Risk Management of Everything: Rethinking the Politics of Uncertainty*, London: Demos.

Power, M. (2007), *Organized Uncertainty: Designing a World of Risk Management*, Oxford: Oxford University Press.

Prakash, A. and Potoski, M. (2006), *Voluntary Environmentalists: Green Clubs, ISO 14001, and Voluntary Environmental Regulations*, New York: Cambridge University Press.

Price, D. J. D. (1963), *Little Science, Big Science*, New York: Columbia University Press.

Price, D. J. D. (1971), 'Some remarks on elitism in information and invisible college phenomenon in science', *Journal of the American Society for Information Science*, **22**(2), 74–75.

Price, D. J. D. and Beaver, D. D. (1966), 'Collaboration in an invisible college', *American Psychologist*, **21**(11), 1011–1013.

Price, D. K. (1965), *The Scientific Estate*, Cambridge, MA: Belknap Press.

Pritchard, C. and Walker, E. (1998), 'Challenges for the enforcement of food safety in Britain', *Food Control*, **9**(1), 61–64.

Quirk, P. J. (1980), 'The Food and Drug Administration', in J. Q. Wilson (ed.), *The Politics of Regulation*, New York: Basic Books, pp. 191–235.

Ragin, C. C. (1989), *The Comparative Method: Moving Beyond Qualitative and Quantitative Strategies*, Berkeley and San Francisco: University of California Press.

Raiffa, H. (1982), 'Science and policy: Their separation and integration in risk analysis', *American Statistician*, **36**(3), 225–231.

Raiffa, H. (2002), 'Decision analysis: A personal account of how it got started and evolved', *Operations Research*, **50**(1), 179–185.

Rawlins, M. (1984), 'Post-marketing surveillance of adverse reactions to drugs', *British Medical Journal*, **288**(6421), 879–880.

Rawlins, M., Fracchia, G. and Rodriguez-Farré, B. (1992), 'EURO-ADR: Pharmacovigilance and research. A European perspective', *Pharmacoepidemiology and Drug Safety*, **1**(2), 261–268.

Rees, N. and Day, M. (2001), 'UK consumption databases relevant to acute exposure assessment', *Food Additives and Contaminants*, **17**(7), 575–581.

Regmi, A., Takeshima, H. and Unnevehr, L. (2008), *Convergence in Global Food Demand and Delivery*, Washington, DC: Economic Research Service, United States Department of Agriculture.

Regulation (EC) No 178/2002 of the European Parliament and of the Council of 28 January 2002 laying down the general principles and requirements of food law, establishing the European Food Safety Authority and laying down procedures in matters of food safety, OJ L 31, 1.2.2002, pp. 1–24

Regulation (EC) No 258/97 of the European Parliament and of the Council of 27 January 1997 concerning novel foods and novel food ingredients, OJ L 43, 14.2.1997, pp. 1–6.

Regulation (EC) No 726/2004 of the European Parliament and of the Council of 31 March 2004 laying down Community procedures for the authorisation and supervision of medicinal products for human and veterinary use and establishing a European Medicines Agency, OJ L 136, 30.4.2004, pp. 1–33.

Reichenbach, H. (1999), 'International food safety and HACCP conference – opening speech', *Food Control*, **10**(4–5), 235–237.

Reinicke, W. H. (1998), *Global Public Policy: Governing without Government*, Washington, DC: Brookings.

Renn, O. (1992), 'Concepts of risk: A classification', in S. Krimsky and D. Gould (eds), *Social Theories of Risk*, Westport, CT: Praeger, pp. 53–79.

Renn, O. (2004), 'The challenge of integrating deliberation and expertise, in T. MacDaniels and M. J. Small (eds), *Risk Analysis and Society: An Interdisciplinary Characterization of the Field*, Cambridge: Cambridge University Press, pp. 289–366.

Renn, O. (2008), *Risk Governance: Coping with Uncertainty in a Complex World*, London: Earthscan/James & James.

Renwick, A., Barlow, S., Hertz-Picciotto, I., Boobis, A., Dybing, E., Edler, L., et al. (2003), 'Risk characterisation of chemicals in food and diet', *Food and Chemical Toxicology*, **41**, 1211–1271.

Rhodes, R. A. W. (2000), 'The governance narrative: Key findings and lessons from the ERC's Whitehall programme', *Public Administration*, **78**(2), 345–363.

Rhodes, R. A. W. (2007), 'Understanding governance: Ten years on', *Organization Studies*, **28**(8), 1243–1264.

Richardson, J. J. (1996), 'Policy-making in the EU. Interests, ideas, and garbage cans of primeval soup', in J. J. Richardson (ed.), *European Union: Power and Policy-Making*, London: Routledge, pp. 3–23.

Rip, A. (1986a), 'Controversies as informal technology assessment', *Science Communication*, **8**(2), 349–371.

Rip, A. (1986b), 'The mutual dependence of risk research and political context', *Science and Technology Studies*, **4**(3), 3–15.

Risse-Kappen, T. (1995), *Bringing Transnational Relations Back In: Non-State Actors, Domestic Structures and International Institutions*, Cambridge: Cambridge University Press.

Risse-Kappen, T. (1996), 'Exploring the nature of the beast: International relations theory and comparative policy analysis meet the European Union', *Journal of Common Market Studies*, **34**(1), 53–80.

Robbins, D. and Johnston, R. (1975), 'The role of cognitive and occupational differentiation in scientific controversies', *Social Studies of Science*, **6**(3–4), 349–368.

Roberfroid, M. B. (1999), 'What is beneficial for health? The concept of functional food', *Food and Chemical Toxicology*, **37**(9–10), 1039–1041.

Rodricks, J. V. (2003), 'What happened to the Red Book's second most important recommendation?', *Human and Ecological Risk Assessment: An International Journal*, **9**(5), 1169–1180.

Roederer-Rynning, C. and Daugbjerg, C. (2010), 'Power, learning or path dependency? Investigating the roots of the European Food Safety Authority', *Public Administration*, **88**(2), 315–330.

Roqueplo, P. (1995), 'Scientific expertise among political powers, administrations and public opinion', *Science and Public Policy*, **22**(3), 175–182.

Rosa, E. A. (1998), 'Metatheoretical foundations for post-normal risk', *Journal of Risk Research*, **1**, 15–44.

Rothstein, H. (2007), 'Talking shops or talking turkey? Institutionalizing consumer representation in risk regulation', *Science, Technology & Human Values*, **32**(5), 582–607.

Rothstein, H., Huber, M. and Gaskell, G. (2006), 'A theory of risk colonization: The spiralling regulatory logics of societal and institutional risk', *Economy and Society*, **35**(1), 91–112.

Rothstein, H., Irwin, A., Yearley, S. and Mc Carthy, E. (1999), 'Regulatory science, Europeanization, and the control of agrochemicals', *Science, Technology & Human Values*, **24**(2), 241–264.

Routledge, P. (1998), '150 years of pharmacovigilance', *The Lancet*, **351**(9110), 1200–1201.

Rowe, W. D. (1977), *An Anatomy of Risk*, New York: Wiley.

Royer, R.-J. (1993), 'Preface – Special Issue: Methodological approaches in pharmacovigilance', *Post Marketing Surveillance*, **7**(1–2), 1–3.

Rushefsky, M. (1982), 'Technical disputes: Why experts disagree', *Review of Policy Research*, **1**(4), 676–685.

Sabatier, P. (1988), 'An advocacy coalition model of policy change and the role of policy-oriented learning therein', *Policy Sciences*, **21**, 129–168.

Sabatier, P. A. and Hunter, S. (1989), 'The incorporation of causal perceptions into models of elite belief systems', *Political Research Quarterly*, **42**(3), 229–261.

Sabel, C. F. and Zeitlin, J. (2010), *Experimentalist Governance in the European Union: Towards a New Architecture*, Oxford and New York: Oxford University Press.

Salter, L. (1985), 'Science and peer review: The Canadian standard-setting experience', *Science, Technology, and Human Values*, **10**(4), 37–46.

Salter, L. (1988), *Mandated Science: Science and Scientists in the Making of Standards*, Dordrecht: Kluwer Academic Publishers.

Sassen, S. (1996), *Losing Control? Sovereignty in an Age of Globalization*, New York: Columbia University Press.

Sassen, S. (2007), *A Sociology of Globalization*, New York: W.W. Norton & Company.

Schilter, B., Andersson, C., Anton, R., Constable, A., Kleiner, J., O'Brien, J., et al. (2003), 'Guidance for the safety assessment of botanicals and botanical preparations for use in food and food supplements', *Food and Chemical Toxicology*, **41**(12), 1625–1649.

Schlich, T. (2004), 'Objectifying uncertainty: History of risk concepts in medicine', *Topoi*, **23**(2), 211–219.

Schmitt, H. (1982), *Pharmacologie*, Paris: Encyclopaedia Universalis.

Schott, T. (1993), 'World science: Globalization of institutions and participation', *Science, Technology, and Human Values*, **18**(2), 196–208.

Schwarz, M. and Thompson, M. (1990), *Divided We Stand: Redefining Politics, Technology and Social Choice*, Philadelphia: University of Pennsylvania Press.

Scott, J. (1998), *Seeing Like a State. How Certain Schemes to Improve the Human Condition Have Failed*, New Haven, CT: Yale University Press.

Selznick, P. (1985), 'Focussing organizational research on regulation', in R. Noll (ed.), *Regulatory Policy and the Social Sciences*, Berkeley & Los Angeles: University of California Press, pp. 363–367.

Shelley, J. H. and Baur, M. P. (1999), 'Paul Martini: The first clinical pharmacologist?', *The Lancet*, **353**(9167), 1870–1873.

Shinn, T. (2005), 'New sources of radical innovation: Research-technologies, transversality and distributed learning in a post-industrial order', *Social Science Information*, **44**(4), 731–764.

Short, J. F. Jr (1984), 'The social fabric at risk: Toward the social transformation of risk analysis', *American Sociological Review*, **49**(6), 711–725.

Short, J. F. Jr (1992), 'Defining, explaining and managing risks', in J. F. Short Jr and L. Clarke (eds), *Organizations, Uncertainties and Risk*, Boulder, San Francisco and Oxford: Westview Press, pp. 3–23.

Shrader-Frechette, K. (1997), 'How some risk frameworks disenfranchise the public', *Risk*, **8**(362), 1–8.

Sinclair, T. J. (2000), 'Reinventing authority: Embedded knowledge networks and the new global finance', *Environment and Planning C: Government and Policy*, **18**(4), 487–502.

Sitter, H. de and Van de Haar, S. (1998), 'Governmental food inspection and HACCP', *Food Control*, **9**(2–3), 131–135.

Skogstad, G. (2005), 'Policy networks and policy communities: Conceptual evolution and governing realities', Annual Meeting of the Canadian Political Science Association, London, Ontario.

Skolbekken, J. A. (1995), 'The risk epidemic in medical journals', *Social Science & Medicine*, **40**(3), 291–305.

Slaughter, A. M. (2004), *A New World Order*, Princeton, NJ: Princeton University Press.

Smith, D. F. and Phillips, J. (2000), 'Food policy and regulation: A multiplicity of actors and experts', in D. F. Smith and J. Phillips (eds), *Food, Science, Policy and Regulation in the Twentieth Century: International and Comparative Perspectives*, London: Routledge, pp. 1–16.

Smith, T. (1992), 'Regulating the food industry in the European Community: Conference report', *Food Control*, **3**(4), 221–224.

Smith-De Waal, C. (1996), 'A consumer view on improving benefit/cost analysis: The case of HACCP and microbial food safety', in *Proceedings of the NE-165 Conference on Strategy and Policy in the Food System*, 20–21 June 1996, Washington, DC.

Snell, E. S. (1986), 'Regulatory decisions and the pharmaceutical industry', *Medical Toxicology*, **1**(Suppl. 1), 130–136.

Sperber, W. (1998a), 'Auditing and verification of food safety and HACCP', *Food Control*, **9**(2–3), 157–162.

Sperber, W. H. (1998b), 'Future developments in food safety and HACCP', *Food Control*, **9**(2/3), 129–130.

Sperber, W. H. (2005), 'HACCP does not work from farm to table', *Food Control*, **16**(6), 511–514.

Stanziani, A. (2005), *Histoire de la qualité alimentaire. XIX–XXe siècles*, Paris: Seuil.

Starr, C. (1969), 'Social benefit versus technological risk', *Science*, **165**(899), 1232–1238.

Starr, C. and Whipple, C. (1980), 'Risks of risk decisions', *Science*, **208**(4448), 1114–1119.

Starr, C. and Whipple, C. (1984), 'A perspective on health and safety risk analysis', *Management Science*, **30**(4), 452–463.

Stephens, M. (1994), 'Commentary on the guidelines for company sponsored safety assessment of marketed medicine (SAMM)', *Pharmacoepidemiology and Drug Safety*, **3**(1), 5–6.

Stephens, T. and Brynner, R. (2001), *Dark Remedy. The Impact of Thalidomide and its Revival as a Vital Medicine*, Cambridge, MA: Perseus Publishing.

Stone, D. (2002), 'Introduction: Global knowledge and advocacy networks', *Global Networks*, **2**(1), 1–12.

Stone, D. (2004), 'Transfer agents and global networks in the "transnationalization" of policy', *Journal of European Public Policy*, **11**(3), 545–566.

Stone, D. (2005), 'Knowledge networks and global policy', in D. Stone and S. Maxwell (eds), *Global Knowledge Networks and International Development: Bridges across Boundaries*, London: Routledge, pp. 89–195.

Tamm-Hallström, K. (2004), *Organizing International Standardization: ISO and the IASC in Quest of Authority*, Cheltenham, UK and Northampton, MA, USA: Edward Elgar Publishing.

Tanner, B. (2000), 'Independent assessment by third-party certification bodies', *Food Control*, **11**(5), 415–417.

Temin, P. (1980), *Taking Your Medicine: Drug Regulation in the United States*, Cambridge, MA: Harvard University Press.

Temin, P. (1985), 'Government actions in times of crisis: Lessons from the history of drug regulation', *Journal of Social History*, **18**(3), 433–438.

Thatcher, M. (1998), 'The development of policy network analyses', *Journal of Theoretical Politics*, **10**(4), 389–416.

Thomas, P. and Newby, M. (1999), 'Estimating the size of the outbreak of new-variant CJD', *British Food Journal*, **101**(1), 44–57.

Thompson, K. M., Deisler, P. F. Jr and Schwing, R. (2005), 'Interdisciplinary vision: The first 25 years of the Society for Risk Analysis (SRA), 1980–2005', *Risk Analysis*, **25**(6), 1333–1386.

Timmermans, S. and Berg, M. (2003), *The Gold Standard: The Challenge of Evidence-Based Medicine and Standardization in Health Care*, Philadelphia, PA: Temple University Press.

Trémolières, J. (2002), *Nomenclature des aliments*, Paris: Encyclopaedia Universalis.

Truhaut, R. (1976), 'Aperçus sur les dangers de l'ère chimique', *Journal de Pharmacie Belge*, **31**, 117–138.

Truhaut, R. (1991), 'The concept of acceptable daily intake: An historical review', *Food Additives and Contaminants*, **8**(2), 151–162.

Turner, S. (2001), 'What is the problem with experts?', *Social Studies of Science*, **31**(1), 123–149.

Unnevehr, L. J. and Jensen, H. H. (1999), 'The economic implications of using HACCP as a food safety regulatory standard', *Food Policy*, **24**(6), 625–635.

Untermann, F. (1999), 'Food safety management and misinterpretation of HACCP', *Food Control*, **10**(3), 161–167.

USDA (2010), *Global Food Markets: Global Food Industry Structure, Economic Research Service Briefing Rooms*, Washington, DC: USDA.

Valverde, J. L., Piqueras García, A. J. and Cabezas López, M. D. (1997), 'La "nouvelle approche" en matière de santé des consommateurs et sécurité alimentaire: La nécessité d'une agence européenne de sécurité des aliments', *Revue du Marché Unique Européen*, **4**, 31–58.

Van Apeldoorn, B. (2000), 'Transnational class agency and European governance: The case of the European Round Table of Industrialists', *New Political Economy*, **5**(2), 157–181.

van Boxtel, C. J. (1993), 'Strategies for PMS of new drugs in the EC', *Pharmacoepidemiology and Drug Safety*, **2**(Suppl. 7), S7–S10.

van den Brandt, P., Voorrips, L., Hertz-Picciotto, I., Shuker, D., Boeing, H., Speijers, G., et al. (2002), 'The contribution of epidemiology', *Food and Chemical Toxicology*, **40**(2–3), 387–424.

van Eijndhoven, J. and Groenewegen, P. (1991), 'The construction of expert advice on health risks', *Social Studies of Science*, **21**(2), 257.

van Schothorst, M. (1990), 'Hazard analysis in hygienic engineering', *Food Control*, **1**(3), 133–136.

van Schothorst, M. and Jongeneel, S. (1992), 'HACCP, product liability and due diligence', *Food Control*, **3**(3), 122–124.

van Schothorst, M. and Jongeneel, S. (1994), 'Line monitoring, HACCP and food safety', *Food Control*, **5**(2), 107–110.

van Zwanenberg, P. and Millstone, E. (2005), *BSE: Risk, Science and Governance*, Oxford: Oxford University Press.

Veggeland, F. and Borgen, S. O. (2005), 'Negotiating international food standards: The World Trade Organization's impact on the Codex Alimentarius Commission', *Governance*, **18**(4), 675–708.

Verdun, A. (1999), 'The role of the Delors Committee in the creation of EMU: An epistemic community?', *Journal of European Public Policy*, **6**(2), 308–328.

Vettorazzi, G. (1987), 'Advances in the safety evaluation of food additives', *Food Additives and Contaminants*, **4**(4), 331–356.

Vitry, A., Lexchin, J., Sasich, L., Dupin-Spriet, T., Reed, T., Bertele, V., et al. (2008), 'Provision of information on regulatory authorities' websites', *Internal Medicine Journal*, **38**(7), 559–567.

Vogel, D. (1986), *National Styles of Regulation*. Ithaca, NY: Cornell University Press.

Vogel, D. (1998), 'The globalization of pharmaceutical regulation', *Governance*, **11**(1), 1–22.

Vogel, D. (2003), 'The hare and the tortoise revisited: The new politics of consumer and environmental regulation in Europe', *British Journal of Political Science*, **33**(4), 557–580.

Wal, J.-M., Hepburn, P., Lea, L. and Crevel, R. (2003), 'Post-market surveillance of GM foods: Applicability and limitations of schemes used with pharmaceuticals and some non-GM novel foods', *Regulatory Toxicology and Pharmacology*, **38**, 98–104.

Waller, P. (2001), 'Pharmacoepidemiology – a tool for public health', *Pharmacoepidemiology and Drug Safety*, **10**(2), 165–172.

Waller, P., Coulson, R. and Wood, S. (1996), 'Regulatory pharmacovigilance in the United Kingdom: Current principles and practice', *Pharmacoepidemiology and Drug Safety*, **5**, 363–375.

Waller, P. C. and Evans, S. J. W. (2003), 'A model for the future conduct of pharmacovigilance', *Pharmacoepidemiology and Drug Safety*, **12**(1), 17–29.

Waller, P. and Lee, E. (1999), 'Responding to drug safety issues', *Pharmacoepidemiology and Drug Safety*, **8**, 535–552.

Walls, I. (1997), 'Use of predictive microbiology in microbial food safety risk assessment', *International Journal of Food Microbiology*, **36** (2–3), 97–102.

Waterton, C. (2005), 'Scientists' conceptions of the boundaries between their own research and policy', *Science and Public Policy*, **32**(6), 435–444.

Weber, M. (1919), 'Wissenschaft als Beruf' and 'Politik als Beruf', Duncker & Humblodt, Munich, reprinted and translated in D. S. Owen, T. B. Strong and R. Livingstone (eds) (2004), *The Vocation Lectures: 'Science as a Vocation', 'Politics as a Vocation'*, London: Hackett Publishing.

Weber, M. (1956), *Wirtschaft und Gesellschaft. Grundriss der verstehenden Soziologie*, Tubingen: Mohr, reprinted in G. Roth and C. Wittich (eds) (1978), *Economy and Society. An Outline of Interpretive Sociology*, Berkeley and Los Angeles: University of California Press.

Webster, C. (1974), 'New light on the invisible college: The social relations of English science in the mid-seventeenth century', *Transactions of the Royal Historical Society*, **24**, 19–42.

Weinberg, A. M. (1972), 'Science and trans-science', *Minerva*, **10**(2), 209–222.

Weinberg, A. M. (1985), 'Science and its limits: The regulator's dilemma', *Issues in Science and Technology*, **2**(1), 59–72.

Weingart, P. (1999), 'Scientific expertise and political accountability: Paradoxes of science in politics', *Science and Public Policy*, **26**(3), 151–161.

Wester, K., Jonnson, A. K., Sigset, O., Druid, H. and Hagg, S. (2008), 'Incidence of fatal adverse drug reactions: A population based study', *British Journal of Clinical Pharmacology*, **65**, 573–579.

WHO (1997), *HACCP – Introducing the Hazard Analysis and Critical Control Point System*, WHO/FSF/FOS/97.2, Geneva: WHO.

WHO (2001), *Joint FAO/WHO Expert Consultation on Risk Assessment of Microbiological Hazards in Foods. Hazard Identification, Exposure Assessment and Hazard Characterization of Campylobacter spp. in Broiler Chickens and Vibrio spp. in Seafood* (No. WHO/SDE/PHE/ FOS/01.4), Geneva: WHO-FAO.

WHO (2002), *The Importance of Pharmacovigilance. Safety Monitoring of Medicinal Products*, Geneva: WHO.

WHO (2007), *Food Safety and Foodborne Illness*, Fact sheet No. 237, consulted at www.who.int/mediacentre/factsheets/fs237/en/

Will, C. M. (2007), 'The alchemy of clinical trials', *BioSocieties*, **2**(01), 85–99.

Winickoff, D. E. and Bushey, D. M. (2010), 'Science and power in global food regulation: The rise of the Codex Alimentarius', *Science, Technology & Human Values*, **35**(3), 356–381.

Wolfe, S. M. (2002), 'Drug safety withdrawals: Who is responsible for notifying patients?', *Pharmacoepidemiology and Drug Safety*, **11**(8), 641–642.

Wray, K. B. (2005), 'Rethinking scientific specialization', *Social Studies of Science*, **35**(1), 151.

Wright, A. C. and Teplitski, M. (2009), 'Thinking beyond the HACCP', *Current Opinion in Biotechnology*, **20**(2), 133–134.

Wuthnow, R. (1979), 'The emergence of modern science and world system theory', *Theory and Society*, **8**(2), 215–243.

Wynne, B. (1980), 'Technology, risk and participation', in J. Conrad (ed.), *Society, Technology and Risk Assessment*, New York: Academic Press, pp. 173–208.

Wynne, B. (1982), *Rationality and Ritual: The Windscale Inquiry and Nuclear Decisions in Britain*, Chalfont St Giles, Bucks: The British Society for the History of Science.

Wynne, B. (1989), 'Establishing the rules of laws: Constructing expert authority', in R. Smith and B. Wynne (eds), *Expert Evidence: Interpreting Science in the Law*, London: Routledge, pp. 23–55.

Wynne, B. (1992), 'Carving out science (and politics) in the regulatory jungle', *Social Studies of Science*, **22**(4), 745–758.

Young, J. H. (1989), *Pure Food: Securing the Federal Food and Drugs Act of 1906*, Princeton, NJ: Princeton University Press.

Zaltman, G. (1974), 'Note on an international invisible college for information exchange', *Journal of the American Society for Information Science*, **25**(2), 113–117.

Zito, A. R. (2001), 'Epistemic communities, collective entrepreneurship and European integration', *Journal of European Public Policy*, **8**(4), 585–603.

Zuccala, A. (2006), 'Modeling the invisible college', *Journal of the American Society for Information Science and Technology*, **57**(2), 152–168.

Index